ELECTRICAL MEASURING INSTRUMENTS

THE ART AND SCIENCE OF ELECTRICAL MEASURING INSTRUMENTS

THE WYKEHAM TECHNOLOGICAL SERIES
for universities and institutes of technology

General Editor:
    J. Thomson, M.A., D.Sc., F.I.E.E., F.Inst.P.
    recently Chief of Technical Studies,
    Sangamo Weston, Ltd.

THIS BOOK is one of a series in which authorities will attempt to introduce young people to the various technologies in the subjects discussed. Primarily these books are aimed at young graduates or apprentices who are about to begin a career in industry or who have just started work. For this reason more attention has been paid to giving a broad view of each subject than to 'necessary and sufficient' mathematical proofs. Indeed the use of mathematics has been sparing.

Another feature of the Technological Series is the mixture of disciplines. For example, when micro-circuits are being discussed, the physical basis of semi-conduction phenomena is inextricably mixed up with the chemistry of almost pure single crystal formation and with the engineering necessary to produce a monolithic circuit. This admixture of disciplines is a feature of industrial research and development. Indeed, it might be argued that this feature of industrial work makes it very unsuitable for the so-called 'specialist' who only recognises his own discipline.

It is hoped both by the authors and by the publishers that this series will open up some new horizons to the young scientist and engineer in industry.

ELECTRICAL MEASURING INSTRUMENTS

E. Handscombe
Product Engineering Manager (Aviation)
Sangamo Weston, Ltd.

WYKEHAM PUBLICATIONS (LONDON) LTD
(A subsidiary of Taylor & Francis Ltd)
LONDON & WINCHESTER
1970

First published 1970 by Wykeham Publications (London) Ltd.

© 1970 The author. All rights reserved. No part of this publication may be reproduced, stored in retrieval system, or transmitted, in any form or by any means, electronic, mechanical, photocopying, recording, or otherwise, without the prior permission of the copyright owner.

Cover illustration—Measuring instruments with accessories

Printed in Great Britain by Taylor & Francis Ltd.
10–14 Macklin Street, London, W.C.2

ISBN 0 85109 130 X

Distribution:

UNITED KINGDOM, EUROPE, MIDDLE EAST AND AFRICA
Chapman & Hall Ltd. (a member of Associated Book Publishers Ltd.), 11 New Fetter Lane, London, E.C.4 and North Way, Andover, Hampshire.

UNITED STATES OF AMERICA, CANADA AND MEXICO
Springer-Verlag New York Inc., 175 Fifth Avenue, New York, New York 10010.

AUSTRALIA AND NEW GUINEA
Hicks Smith & Sons Pty., Ltd., 301 Kent Street, Sydney, N.S.W. 2000.

NEW ZEALAND AND FIJI
Hicks Smith & Sons Ltd., 238 Wakefield Street, Wellington.

ALL OTHER TERRITORIES
Taylor & Francis Ltd., 10–14 Macklin Street, London, W.C.2.

AUTHOR'S PREFACE

This book has been written to give the young engineer some idea of the problems which will face him in designing, making or using electronic instruments.

It has been pointed out in many places throughout this volume that the designing of electrical measuring instruments is as much an art as a science and I suspect that the same could be said of a large number of other engineering products.

As the reader will be quick to realise, much of this book owes its origin to the firm of Sangamo Weston, Ltd. To the Chairman of this Company and to its Chief Engineer my warmest thanks go out.

Their co-operation has also made possible the publication of many photographs of important types of instruments.

February, 1970. E. HANDSCOMBE.

CONTENTS

Chapter 1	INTRODUCTORY	1
Chapter 2	BRIEF HISTORY OF ELECTRICAL MEASURING INSTRUMENTS	5
Chapter 3	CLASSIFICATION OF INSTRUMENT TYPES AND THEIR USES	9
Chapter 4	DESIGN	39
Chapter 5	MATERIALS AND FINISHES	54
Chapter 6	PRESENTATION	74
Chapter 7	MANUFACTURING TECHNIQUES	81
Chapter 8	EXTENSION OF RANGE	88
Chapter 9	THE SPECIAL PROBLEMS OF AIRCRAFT INSTRUMENTS	97
Chapter 10	TRANSDUCERS	104
Chapter 11	WHAT OF THE FUTURE?	111
Index		114

CHAPTER 1
introductory

THE following statement, attributed to Lord Kelvin, applies to all branches of engineering and science and not least to the subject of this volume.

' I often say that when you can measure what you are speaking about and can express it in numbers, you know something about it; but when you cannot measure it, when you cannot express it in numbers, your knowledge is of a meagre and unsatisfactory kind; it may be the beginning of knowledge but you have scarcely in your thoughts advanced to the stage of science whatever the matter may be.'

The startling progress of the whole electrical industry from almost nothing to its present size and complexity in little more than a 100 years is strongly linked with the establishment of the science of electrical measurement. Until the invention and development of suitable measuring devices most electrical phenomena, although known for many years, were but little understood and remained scientific curiosities with little or no thought given to their possible commercial application.

One of the earliest pressures which led to the development of practical measuring instruments came from the rapidly developing electrical supply industry which had its beginnings in the early 1850's. The first practical alternating current distribution system was demonstrated in 1886 and in 1888 the Weston Electric Instrument Company was formed by Dr. Edward Weston.

The use of these early instruments undoubtedly contributed greatly to the development of the electrical industry and, as the performance of dynamos, motors, and generators was more fully understood, progress to improved versions became ever more rapid.

The rapid development in electrical machinery also gave rise to a demand for more and better electrical measuring instruments so that the advances in these two new industries in their early days were very much interlinked.

As instruments which had primarily been developed at the instigation of the electricity supply industry became more readily available they were increasingly applied to the problems of other new industries such as telephony, telegraphy and illumination which were also making great strides.

The ability to measure and hence control electrical quantities has undoubtedly contributed greatly to the development of today's complex electrical and electronic industry.

During research and development of most electrical devices accurate electrical measurement plays a very important part and, in spite of the development of highly accurate measuring devices such as the digital voltmeter, the humble indicating instrument is still an essential piece of equipment in any electrical laboratory.

In many production processes measuring instruments play a vital part in enabling the human operator to monitor the process continuously and hence control it within the prescribed limits.

Even the use of such complicated pieces of machinery as the modern jet aircraft would be unthinkable were it not for the information presented to the pilot by the indications of electrical measuring instruments.

Examples of the use of measuring instruments in modern technology are endless and the modern instrument industry continues to strive for better measuring devices to offer the user.

Measurement: the beginning of control

In most fields of scientific endeavour before any phenomenon can be understood it is necessary to make measurements of it. Where such phenomena or their effects are electrical a wide range of measurements are carried out using the types of electrical indicating instruments which are the subject of this volume.

With suitable ancillary apparatus these measurements may range from, for example, the output of a power station alternator to the minute voltages produced by thermocouples.

The purpose of measurement is usually twofold; first to obtain a better understanding of the relationship between cause and effect and secondly to facilitate the development of an effective means of control.

To give but one simple example of this principle in practice, imagine the chaos which would ensue if the voltage of the power supply to our homes could not be accurately measured and hence controlled.

How many of the material advantages of our technological age would we be able to enjoy if no means of electrical measurement had ever been discovered? I venture to suggest that many of the more obvious electrical things such as television would never have been invented and many of the devices with less obviously electrical dependence such as motor cars would at the very least not have progressed as rapidly.

Many of the complex automatic processes used in industry owe their very existence to the hosts of measurements which were made on earlier and simpler pilot plants. These measurements, both electrical and non-electrical, allowed the scientists and technicians to forecast the performance of and devise automatic control systems for the present large scale installations in such industries as oil and plastics. Were it not for the science of measurement these and many other industries would not have progressed beyond the craftsman stage, which relied largely on the intuitive control exercised by individual workers with long experience. Measurement and subsequent control have led to processes being carried out on a scale which would be unthinkable if control were left to the unaided intervention of individual craftsmen.

The reader will doubtless be able to think of many examples for himself which illustrate the way in which measurement has been the vital beginning in the progress to partially or fully automatic control.

Measurement: the yardstick of progress

Progress, particularly in scientific fields, is closely linked with knowledge of the facts. Theories however plausible can ultimately only be accepted when they are demonstrated to fit the facts. Scientific facts are largely a matter of measurement, so that the progress in any branch of science is closely linked to the ability to measure the parameters with which the particular branch is concerned. As the quality of measurement progresses, new and unexpected effects frequently come to light and lead to yet further progress.

Man's progress from the earliest times has been linked with his awareness of his environment, his ability to adapt himself to that environment and to make use of the materials available to him. Awareness of his surroundings in the earliest times was closely related to his ability to measure such parameters as distance, speed, time, weight, heat, cold, etc., in order to reason and communicate with his fellows concerning the best manner in which to utilize the material resources.

Whilst modern man has no longer any need to concern himself unduly with these more primitive considerations he is still limited by his ability to understand his surroundings, which again points to the necessity for measurement. If his preoccupation is with any form of science then the measurements required will be scientific and undoubtedly many of these will be carried out with the aid of one form or another of electrical measuring apparatus.

Measurement: the basis of standards

In order for one scientist to communicate adequately with another it is obviously desirable that they should be able to define their findings in terms of universally understood and accepted units.

Early experimenters in the area of electrical phenomena had no names for their units and magnitude was defined only in extensions of the C.G.S. system which had originally been used for expressing mechanical quantities. One system was developed for electrostatics and a different one for magnetostatics. This led to considerable confusion since certain units could be defined in both systems with completely different results. As the electrical industry developed it was seen to be imperative that units should be named and also have values such that commonly occurring quantities could be expressed in simple numbers. Names of the units were chosen from the names of famous investigators such as ohm, farad, henry, volt, ampere, etc. and were given values which were conveniently related to the C.G.S. magnetostatic system with factors which were whole numbers.

Since that time various national and international bodies have sought to clarify the definitions and to get universal acceptance of their magnitude. Collaboration between national standardizing bodies such as the National Physical Laboratory in Britain and the Bureau of Standards in America has resulted in universal agreement on the magnitude of most electrical units to within a few parts in a million. This means, for example, that anywhere in the world for all practical purposes a current of 1 ampere has the same magnitude. This result has only been achieved by the making of enormous numbers of the most careful measurements it is possible to make. The process is a continuing one with every effort being made to reduce even these minute discrepancies.

CHAPTER 2
brief history of electrical measuring instruments

The properties of natural magnets have been mentioned in literature as far back as 600 B.C. and the word 'magnet' is thought to have been derived from the province of Magnesia where natural magnets were found by the Greeks. Plato also discussed natural magnets about 400 B.C. and the Chinese described their use as crude compasses for navigation about A.D. 1100. Long magnets pivoted in the middle were found to point always north and south and so gave rise to the naming of magnet poles as 'north seeking' or north poles and 'south seeking' or south poles. It should be remembered, however, that these are rather confusing terms since the Earth's magnetic pole situated near the geographical north pole is in magnetic terms a south pole.

The first serious scientific study of magnets is recorded in a book by William Gilbert written about 1600. It is concerned entirely with natural magnets although it was noted that bars of iron could be magnetized by stroking them with a natural magnet.

Coulomb established a law for the forces between concentrated poles but it was not until 1820 when Hans Christian Oersted discovered that a wire carrying a current exerts a force on a magnetized needle that the connection between electrical and magnetic effects was realized and the foundation for much of the modern instrument industry was laid.

Oersted's discovery stimulated the French physicist André Marie Ampère to carry out experiments of his own and he rapidly discovered the further fact that wires carrying currents in the same direction attracted each other. From there he went on to establish the law that the force F of attraction or repulsion between two wires carrying currents of I_1 and I_2 is given by the expression $F = \mu I_1 I_2 / d$ where d is the distance between them.

The first true instrument to utilize the magnetic effect of a current was that produced by the German physicist Johan Schweigger about 1822 and consisted of a rectangular coil inside which a magnetic needle was suspended. It suffered from the disadvantage that the controlling force was the Earth's magnetic field and various attempts to overcome this were made by placing stronger magnets in the vicinity of the device. In 1858 Sir William Thomson, who later became Lord Kelvin, added a mirror system and by the use of additional coils made an astatic version thus eliminating the effects of the Earth's field .Various slightly different forms of galvanometer based on the moving-needle principle were developed by workers such as Pouillet and Von Helmholtz but it was not until 1879 that the first portable direct reading ammeter was produced by Ayrton and Perry. This version had a large permanent magnet to provide the controlling field and was therefore largely inependent of the Earth's field. This instrument is illustrated diagrammatically in fig. 2.1.

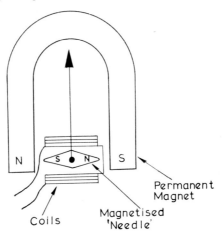

Fig. 2.1.

In 1821 Michael Faraday had by his own independent experiments discovered the interaction between a coil carrying a current and a magnetized needle. He also found that a conductor carrying a current would rotate continuously around a magnet pole if allowed to do so. This discovery opened the way to the invention of the permanent magnet moving coil instrument but in fact the principle was not exploited until 1836 when William Sturgeon made a moving coil galvanometer. His version suspended a coil in the field of a permanent magnet but did not employ an iron core inside the coil to strengthen the field. This important addition was first made by Lord Kelvin in 1867 for his *Syphon* Recorder, but it was not until 1882 that d'Arsonval used the moving coil in a mirror galvanometer. This is illustrated in fig. 2.2 and led in 1888 to the production by Dr. Edward Weston of the first direct reading portable instrument of this type. This instrument was the true ancestor of all the modern permanent magnet moving coil instruments and it is shown in fig. 2.3.

Since that time innumerable variations, amendments and improvements have been made to moving coil instruments. Their originators are largely anonymous and it is interesting to note that Dr. Weston's basic design remains unchanged except for those features which have been improved solely as the outcome of the availability of better materials than he had at his disposal.

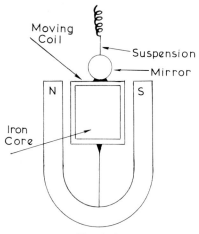

Fig. 2.2.

The moving iron instrument also owes its invention to a fact discovered about 1820 that a coil carrying a current magnetizes iron rods placed near it and also attracts them. This gave rise to various forms of attraction and repulsion types of moving iron instrumen associated with such names as Lord Kelvin and Nalder.

Ampère's discoveries of 1820 also led to the development of the style of instruments known as dynamometers and again Lord Kelvin amongst others contributed to the various developments. The first dynamometer wattmeter was constructed by Ayrton and Perry in 1881 and was made available as a commercial instrument by Siemens in 1884.

Electrostatic instruments were made as crude electroscopes and electrometers by various experiments but once again it was Lord Kelvin who made the first really practical instrument when in 1856 he devised

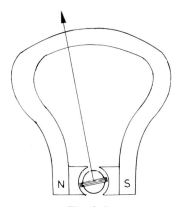

Fig. 2.3.

the quadrant electrometer which is named after him. This was not made available as a commercial instrument until the later 1880's and Kelvin improved the sensitivity of the device in 1890 by making it multicellular.

No basic new principle for an indicating instrument has really been discovered since that time, even the moving magnet instrument which has aroused interest from time to time is really only an improved version of the original moving needle instrument made possible by advances in magnet materials.

CHAPTER 3
classification of instrument types and their uses

IN most common types of electrical indicating instruments, there are two distinct forces acting upon the moving system. They are:
(1) A deflecting force.
(2) A restoring force.

In addition, there is usually a third force which is a damping force, and it is the way in which these forces are developed and controlled, which is the primary concern of the instrument design engineer. The manner in which the deflecting force is developed, also classifies the instrument into its appropriate type.

The main types of instruments with the essential features of each will be found under the following headings.

1. *Permanent magnet moving coil*

The moving coil instrument is by far the most commonly used electrical measuring instrument and, when considered in conjunction with various accessories such as rectifiers, thermocouples, etc., is extremely versatile.

It consists, as its name implies, of a coil of wire, usually copper, suspended by a suitable bearing system, in such a manner, that it is free to rotate. The coil is suspended in the field produced by a permanent magnet and when a current is passed through the coil the interaction occurring between the flux produced by that current and the flux of the permanent magnet causes the coil to rotate, until the torque produced is balanced by a restraining force, which is usually supplied by a spring or springs.

It is essentially a current sensitive device, and since the torque is dependent on the non-varying flux from the permanent magnet, will only give a steady indication on direct current.

The earliest versions relied on gravity as the restoring force, but this imposed serious position errors and non-linearity, due to the sine law which relates force to deflection. This is illustrated diagrammatically in fig. 3.1 from which it will be seen that the restraining or restoring torque is $Wl \sin \theta$ where W is the control weight, l its distance from the centre of rotation and θ the angle through which it is deflected.

The limitations of the gravity control system led to the development of hairsprings for use as the restoring force and these springs also form a convenient path to lead the current to and from the moving coil.

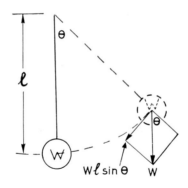

Fig. 3.1.

In order to exploit the non-gravity sensitive restoring force provided by hairsprings, it is also necessary for the centre of gravity of the moving system to be coincident with the axis of rotation. This is usually achieved by means of balance weights attached to balance arms on the moving system. The weights are adjustable in their distance from the axis of rotation and are adjusted until the instrument indication is unaffected by the instrument position. A common arrangement of balance weights is shown in fig. 3.2.

It is usual, although not essential, for the balance arms to be arranged

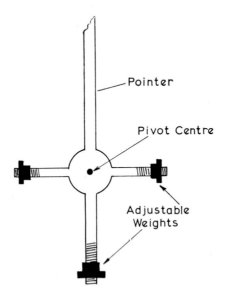

Fig. 3.2.

at right angles to each other, and also at right angles to the axis of rotation.

The exact design of the balancing system is governed by a number of features, such as the design unbalance of the moving system caused by such items as the pointer, and the likely variation from one instrument to another, due to unavoidable variations in piece parts.

In consideration of the design of a balancing system, it is necessary to take into account the effects on other performance parameters due to the weights. They add to the stress on the bearings by increasing the 'dead' weight, and also may contribute significantly to the 'moment of inertia' of the moving system.

Many different systems of balance weight adjustment have been used by different manufacturers, such as threaded weights, weights fixed by adhesives, solder and various ingenious variations of one or more of these methods. All the methods used, however, have the same ultimate aims of ease of adjustment and permanence after adjustment.

Magnetic circuits

The magnetic circuits of permanent magnet moving coil instruments fall broadly into two classes. First, the external magnet design in which the major portion of the magnetic circuit is outside the area swept by the moving coil, and secondly, the internal magnet or centrepole design made practical by the higher energy and high coercive forces available from modern magnet materials. In this version, the magnet is located inside the moving coil, thus permitting the design of very compact mechanisms.

We will consider the external magnet design first. The normal prerequisite of any magnetic circuit for a moving coil instrument is that it shall produce a high, radial, evenly distributed magnetic flux in which the coil rotates. An evenly distributed field is required in order that the instrument scale shall be essentially linear and the higher the flux the more sensitive the instrument can be made.

The essential parts of the magnetic circuit are:

(*a*) the magnet, which is often made in the shape of a ' C ', (*b*) the soft iron polepieces surrounding the coil, and (*c*) the soft iron core inside the coil. Early magnet materials such as the chrome and tungsten steels had very low coercivities, therefore, it was necessary to have long magnets, often horseshoe shaped, in order to reduce the self demagnetizing effects. As magnet materials were improved, magnets were progressively shortened, and reduced in volume and the ' C ' shape is often substituted in modern instruments by short square or rectangular section block magnets; the magnetic circuit being completed by suitable soft iron pole shoes. One interesting variation of this design uses a composite sintering, incorporating both magnet and soft iron pole shoes in one piece. This design of magnet, whilst quite effective, is inclined to be rather expensive to produce, due to the very close control required to give consistent performance.

A diagram showing the general layout of the external magnet circuit is shown in fig. 3.3.

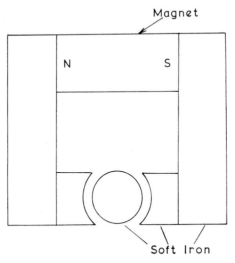

Fig. 3.3.

Although for most moving coil instruments the aim is to achieve a linear current/deflection characteristic, it is sometimes required that the relationship shall conform to a non-linear law. For example, for a dB meter, in order to obtain a linear dB scale, it is necessary to provide a non-linear flux distribution. This can be achieved by shaping the pole-pieces or the soft iron core or even, in some cases, both parts.

The problems of designing the shapes required are not, unfortunately, amenable to exact calculation, especially if the shaping required is drastic. This is due to fringing and leakage which are difficult, if not impossible, to calculate exactly. However, various design techniques are available, which will give a good approximation to the required shape. It is usual after the first calculation to make subsequent small modifications by trial and error and only long experience will reduce this rather tedious procedure to a minimum.

One disadvantage of the external magnet design described, is its tendency to have an external leakage field which makes it susceptible to the proximity of magnetic objects such as steel panels and also it is liable to be affected by stray fields from other sources. The undesirable effects of this leakage can be minimized by careful design in avoiding sharp changes of flux direction, by ensuring minimum gaps between mating parts of the magnetic circuit and by the provision of suitable soft iron shields on the inside or outside of the instrument case.

This disadvantage is largely overcome by the internal magnet design which has the additional advantage of smaller size, where space is at a premium. The layout of the magnetic circuit for this type of mechanism is shown in fig. 3.4. Since the magnet is totally surrounded by a soft iron return ring, the likelihood of leakage flux is much reduced, and the whole magnetic circuit is of a much 'tighter' design. Due to the restricted magnet volume, this type of mechanism is not usually employed where extreme sensitivity is required as the maximum gap flux which can be achieved is always less than can be obtained with an external magnet if space presents no limitation.

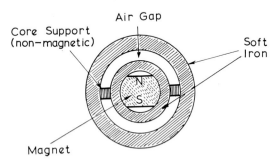

Fig. 3.4.

An interesting design has recently been introduced by one company wherein the moving element assembly can be used with a conventional internal magnet, but by the simple removal of the soft iron return ring it can be replaced by an additional external magnet unit which results in nearly 100% increase in gap flux. The design of the external magnet is also such as to be almost completely self-shielding so the advantage of the normal internal magnet in this respect is not lost. A view of this mechanism showing both internal and external magnets is shown in fig. 3.5.

Moving coil instruments are made in a considerable range of sizes

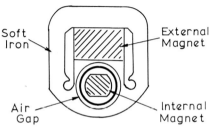

Fig. 3.5.

and accuracies starting from laboratory standards with a scale length of about 12 in. and an accuracy of $\mp 0.1\%$ or better, to instruments with scale lengths under an inch and accuracies of only $\mp 5\%$ or so.

Available sensitivities range from a few μA full scale to some tens of mA. It is not usual for the basic movement sensitivity to be greater than say 50 mA since the number of turns required on the moving coil would be very low, and also because normal hairsprings are not suitable for carrying currents higher than this. The range, however, can be extended upwards almost indefinitely by the use of shunts. The addition of suitable series resistors will also make the device into a voltmeter.

The moving coil instrument is used for an extremely wide range of measurements. Its primary use is on direct current where it can be made to measure from a few μA to many amperes full scale, and from a few mV up to many kV. By the use of rectifiers or thermocouples an almost equally wide range of alternating current measurements may be made. More information on rectifiers and other devices is given in a later chapter. The combination of a moving coil mechanism with appropriate shunt and series resistors, rectifiers, current transformer, etc. and suitable switches in a single unit is the basis of all the well-known multimeters, so widely used in electrical and electronic laboratories.

The selection of an instrument for a particular measurement or series of measurements, demands some knowledge on the part of the user and the following comments on this aspect of the problem may be found helpful. In very specialized or unusual applications, the instrument manufacturer's advice should be sought. Provided he is given full information concerning the problem, the manufacturer will almost always be found willing to help, since his good name can all too easily be damaged by the poor performance of his product in an unsuitable application.

The selection of an instrument for a particular function should always be linked to the quality of information which it is hoped to obtain from the measurement. A careful assessment should be made initially of the required accuracy, bearing in mind that instrument accuracies are usually quoted as a percentage of full scale deflection, so that an instrument of nominal accuracy of 1% of F.S.D. may have an error of 2% of indication when used at half scale. This assessment of required accuracy is probably the most critical factor in selecting the right instrument for the job and great care should be exercised, neither to set too high a standard which will have to be paid for in size or price or both, nor too low a standard which will render the measurement either useless, or even more important, misleading. British Standard 89 gives a guide to the accuracy available from good commercial quality instruments. In certain instances these accuracies may be improved upon, but usually at an enhanced price. Before deciding the accuracy required, some consideration should be given to the conditions under which the instrument will be read. Will the instrument be well lit? Will the user be able to position himself in the best place to read it accurately? Will the instrument be likely to be affected by adverse conditions of temperature or vibration? A guide to the effects of temperature on instruments can also be found in B.S. 89.

Consideration of these, and any other relevant factors will enable a decision to be made as to the required accuracy. This will have a considerable bearing on the necessary instrument size, since obviously the higher accuracies are only available in the larger instruments.

Having settled the accuracy and size of the instrument, the choice of any particular manufacturer's model involves considerations outside the scope of this book, but at least two further basic points must be taken into account in this choice. First, will the instrument be required to function under particularly adverse industrial or service conditions such as would necessitate a sealed case? If not, will it be subjected to solvents or other chemicals which are liable to affect many of the modern injection moulded acrylic cases? Secondly, some account must be taken of the possible effect of the instrument itself on the measurement. Since all instruments absorb some power, this power must be supplied from the measurement circuit and particularly in low power circuits, this power drain may have an adverse effect, unless taken into account in the first place. The significant resistance parameters can be obtained from the catalogue and usually only a simple calculation is necessary to determine, say, in the case of a milliammeter or ammeter, whether the volt drop is significant in the particular circuit, or whether in the case of a voltmeter the current drain taken by the instrument is likely to disturb the true accuracy of the measurement.

Finally it should be remembered that precision grade instruments are normally only suitable for use in a horizontal position, whereas, industrial grade panel instruments will normally have been made so as to achieve their stated accuracy when mounted in a vertical position, and the manufacturer should be given full information at the time of ordering, if the conditions of use do not comply with this simple convention.

An interesting and very useful variation of the permanent magnet moving coil instrument is the permanent magnet moving coil ratiometer. As its name implies its purpose is to measure ratio, in fact the ratio between two currents. Its general construction is often similar to the ordinary moving coil instrument, but it has two moving coils mechanically joined, either side by side, or one above the other, rotating on a common centre. The magnetic circuit is arranged, however, by appropriate shaping so that at any particular deflection the two coils experience different magnetic fluxes. The torques of the two coils are arranged to oppose each other and a steady reading is obtained when $N_1 I_1 \phi_1 = N_2 I_2 \phi_2$ if N_1 and N_2 are the number of turns, I_1 and I_2 the currents and ϕ_1 and ϕ_2 the fluxes concerned with coil number 1 and coil number 2 respectively.

Instead of hairsprings the currents are led to and from the coil by very fine metallic ligaments whose size, length and shape are designed to provide the minimum possible restraining force. If this has been done effectively, the instrument will measure current ratio independent of actual current over quite a wide range. Due to the limitations of the special construction, it is not usually possible to measure ratio between currents in the low μA region, but if currents of half a mA or above are available, a perfectly satisfactory instrument can be made. One of the commonest uses of this type of instrument is in the measurement of resistance, such as is necessary, for example, in resistance thermometry. A typical instrument layout is shown in fig. 3.6 and a simple circuit for

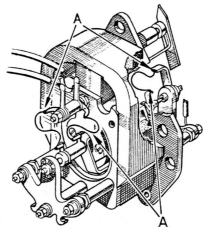

Fig. 3.6.

measurement of say a platinum resistance thermometer in fig. 3.7. The particular advantage of the ratiometer in such a circuit is that it is largely independent of the currents in the two limbs and hence of the supply voltage.

Many ingenious circuits have been designed for ratiometers to enable them to be used for the measurement of frequency, power factor etc. In addition, in conjunction with suitable transducers, they are used to indicate pressure, position, or in fact, any parameter which can be translated into a change in current or resistance ratio.

One of their common applications is as aircraft indicators for the measurement of pressure, position and temperature where their independence of supply voltage is an especially attractive feature. They are, however, because of their construction, comparatively expensive to produce, demanding a fairly high degree of manual skill. The advent of small, cheap and reliable semiconductor devices, such as transistors

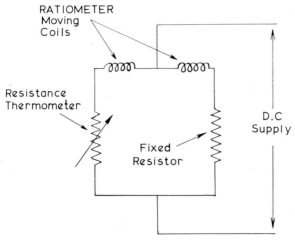

Fig. 3.7.

and zener diodes is tending to lead to the replacement of ratiometers by stabilized voltage networks and conventional moving coil indicators.

Moving iron instruments

Moving iron instruments may be broadly classified into two types, (1) attraction type and (2) repulsion type.

In both types the current to be measured or a current proportional to the voltage to be measured is passed through a field coil consisting of a number of turns of copper wire. It is usual for any particular design of instrument to maintain a consistent value of ampere turns for full scale deflection regardless of range, which means that the field coil may consist of many turns of wire for low current ranges and down to one or two turns for the highest current ranges. A typical design figure for full scale deflection would be 100 ampere turns.

In the attraction type instrument the field produced by the coil, when current is passed, is used to attract a piece of soft iron which is attached to a pivoted staff carrying the pointer. The deflection of the iron is restrained by the usual instrument type hairspring. By suitable design of the shape of the field coil and the iron, the resulting current/deflection relationship can be controlled to give a reasonably linear scale shape over the upper two-thirds of the scale.

In the repulsion type of instrument two pieces of soft iron, one fixed and the other movable, are placed inside the field coil, the movable one being attached to the pivotted staff carrying the pointer. As before, the restraining or restoring force is provided by a hairspring, also attached to the same staff.

In both cases the force acting on the movable iron is independent of the polarity of the energizing current in the field coil so that the instruments will indicate on both d.c. and a.c.

Typical examples of the two basic types of moving iron instrument are illustrated in figs. 3.8 and 3.9 and details of the torque equations are given in a later chapter.

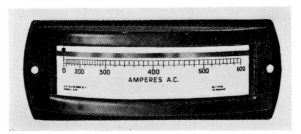

Fig. 3.8.

Both types of moving iron instrument are liable to suffer from certain errors which, unless minimized by careful design, may lead to seriously incorrect readings under certain conditions:

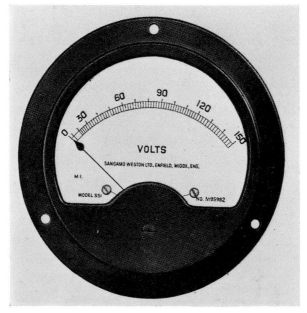

Fig. 3.9.

1. *Hysteresis errors*

Owing to hysteresis in the irons these instruments have a tendency to read differently on increasing current, as opposed to decreasing current, the tendency being to read higher on decreasing current. This error may be reduced to a minimum by first keeping the flux density in the iron as low as is practicable consistent with producing an adequate operating torque and secondly, by using irons made of material such as radiometal, which has an inherently low hysteretic effect. By a combination of these two means the hysteresis can be reduced to a negligible amount, say less than 0·2% difference between readings taken with increasing and decreasing currents on a typical panel instrument.

Due to the low operating flux, moving iron mechanisms are also very susceptible to disturbance by stray magnetic fields, and it is virtually essential for the mechanisms to be shielded. The proper design of shield also needs some care as the proximity of the additional magnetic material outside the field coil can lead to further errors due to the unwanted flux linkage. On d.c. they can contribute to the hysteresis or reversal error and on a.c. particularly with increasing frequency may cause large errors due to added iron losses.

2. *Frequency errors*

The above problems of iron losses are one of the major contributory factors in the tendency of moving iron instruments to have an increasing error with the frequency of the applied current. Their low frequency usefulness is only limited by the frequency at which a steady reading can be obtained. A typical panel instrument could be used down to about 20 to 30 Hz. As the frequency is increased the errors also increase and may become quite considerable above about 100 Hz, unless steps are taken to eliminate them. The errors which occur affect ammeters and voltmeters differently, so they will be considered separately under these headings.

Frequency errors of moving iron ammeters

The major frequency errors in moving iron ammeters are due to the losses in the two operating irons, and the shielding. As previously suggested in the case of the operating irons these errors can be minimized, although not eliminated, by careful selection of material, and by working the material towards the lower end of its B.H. curve. The losses in the shield can also be minimized by material selection, and by keeping the shield as far away from the field coil as possible. It is also important to design the shield so that circulating currents in it are reduced to a minimum, and for this reason it is the usual practice to machine narrow slots in the shield to lengthen the paths and hence increase the resistance. By these means it is possible to make ammeters which have only small errors up to several hundred Hz. Depending on the current range and hence the size of the wire used to wind the field coil, additional errors may arise above this frequency, due to skin effects in the wire, the larger diameters giving rise to measurable errors above about 1000 HZ. It is not usual to attempt any form of frequency compensation in ammeters, but it is quite possible to calibrate them for use over a narrow band of frequencies anywhere in the range up to several thousand Hz. It must be remembered, however, that outside the narrow specified band the errors may well be considerable and bear no relationship to the d.c. or power frequency calibration.

Frequency errors of moving iron voltmeters

In addition to the errors in ammeters caused by iron losses the voltmeter suffers from a further error due to the inductance of the field coil. This results in the impedance of the voltmeter containing a term which is dependent on frequency thus:

If L is the inductance of the field coil in henries and $\omega = 2\pi \times$ frequency in Hz and R = series resistance then Z (the impedance) = $\sqrt{(R^2 + \omega^2 L^2)}$ or $R + j\omega L$. The significance of this error will obviously vary with the voltage range of the instrument since, if it is assumed that same field coil is used for all voltmeters, then the higher the range the more the effect of the inductive reactance will be swamped by the series resistance.

A good measure of compensation for all the errors involved can be obtained by shunting the series resistor with an appropriate capacitor. The exact value of this capacitor is not amenable to calculation, due to the various effects which contribute to the total error, so it is usual to derive an initial value by assuming that the main error is due to the inductive reactance and using the approximate expression $C = L/R^2$. By using this value initially the exact capacitance required is then found by experiment, measuring the instrument errors at different frequencies and adjusting the capacitance as appropriate. It is not possible to obtain perfect compensation except over a limited frequency range, so it is usual to make the best compromise by splitting the errors over the range required. It will also be found that where the best accuracy is required each instrument will have to have its compensation individually adjusted. To save having a variety of capacitor values the proportion of R which is shunted, may be varied slightly from instrument to instrument.

Scale-shaped moving iron

The usual design aim for moving iron instruments is to maintain a reasonably linear current deflection relationship over the major portion of the scale. This can be achieved in both attraction and repulsion types by iron shaping and by careful positioning of the fixed and moving irons respectively.

There is, however, quite a demand for a special version of the instrument which has a reasonably linear scale up to about 70% deflection and above this a violently compressed scale. A typical requirement would be 10 amperes at 70 to 75% deflection, and full scale about 50 amperes. One application for this version of the instrument is to monitor the current taken by direct on-line starting motors. The range called for is such that the initial starting surge does not take the instrument beyond full scale, whilst the running current is in the more linear portion of the scale.

One method of achieving this requirement is with a repulsion type instrument in which a second fixed iron is added in such a position that it repels the moving iron strongly as the deflection approaches full scale. The determination of the necessary size, shape and position of this additional fixed iron is a matter for experiment rather than calculation, particularly since its introduction may easily cause parasitic torques at right angles to the deflection torque. Fortunately, the ' overload ' part of the scale is not normally required to have any great precision.

All that has been said about moving iron instruments applies particularly to those having deflection angles up to about 100°. Most of it is equally applicable to designs having deflection angles greater than this up to about 270° and a number of ingenious designs have been made by different manufacturers to obtain these larger angles. The detailed consideration of these designs, their method of operation and possible errors are beyond the scope of this volume, but generally all the same principles apply, although their explanation in some instances is rather complex.

Moving iron instruments are made in a range of sizes from those having scale lengths of an inch or so up to those with a scale length of 12 in. or more. They are generally not as accurate as the moving coil instrument, but they have the advantage of being suitable for d.c. or a.c. and in the latter case of measuring true R.M.S. values. They will therefore indicate correctly on distorted waveforms provided that the ' crest factor ' is not sufficient to cause the irons to run near or into saturation. Most commercial designs will work satisfactorily with ' crest factors ' not exceeding about 2, but in case of doubt the manufacturer's advice should be sought.

The method of damping employed almost universally is pneumatic, with vanes or pistons attached to the moving system, running in closely fitting damping chambers or cylinders. A few designs have been made employing eddy current damping, but this presents difficulty as the presence of a magnet is liable to affect the deflection characteristics. More recently designs have appeared using viscous damping based on the use of high viscosity fluids such as the silicone oils, which are chosen for their limited change of viscosity with temperature and also because of the wide range of available viscosities.

The moving iron instrument is suitable for many measurement

applications where accuracy is not of supreme importance. It finds its most useful role in the measurement of voltage and current at power frequencies where its true R.M.S. reading makes it especially useful if the waveform is at all doubtful. Generally its high current consumption as a voltmeter and its high voltage drop as an ammeter do not preclude its adoption in the power industry where it is widely used as a cheap and reliable indicator. These factors, however, do present pitfalls for the unwary in the lighter current applications, which together with its lack of sensitivity cause it to be generally viewed with disfavour by most of the electronics and light current side of the electrical industry.

The air cored dynamometer

The electrodynamic moving coil instrument or dynamometer can be likened in some ways to a permanent magnet moving coil instrument in which the permanent magnet and its magnetic circuit have been replaced by field coils. It is commonly made in three forms:

(1) as a voltmeter, when its field coils are connected in series with the moving coil;
(2) as an ammeter when its field coils carry the whole current and its moving coil is connected across a shunt;
(3) as a wattmeter when the field coils and moving coils are brought out separately.

Typical circuit diagrams for these three variants of the dynamometer are shown in fig. 3.10 and a photograph of a typical construction in fig. 3.11.

The two fixed coils are mounted one on either side of the moving coil. They are wound in the shape of an annulus and positioned as close as possible to the moving coil and of such a diameter that it swings inside them when deflected. The moving coil also is circular, often with slightly flattened ends. As far as possible metal is eliminated from close proximity to both fixed and moving coils so that no unwanted eddy currents are induced when the coils are energized by alternating current. The restoring force for the moving coil is provided by a pair of hairsprings which also form the connections to and from the moving coil. As a matter of constructional convenience the two springs are often mounted one above the other at one end of the moving coil. The moving coil is carried on a light alloy staff mounted in jewelled bearings, the lower end of the staff carrying vanes which work in damping chambers.

The two fixed coils for any particular instrument are made as nearly identical to each other as possible and during assembly the position of each coil relative to the moving coil is carefully adjusted so that each coil will contribute half the flux encompassed by the moving coil. This adjustment helps to ensure a consistent scale shape and, as will be seen

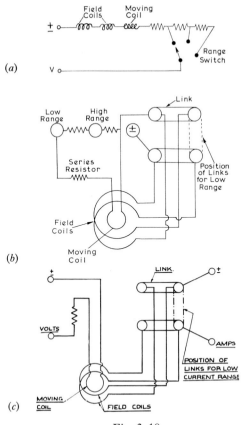

Fig. 3.10.

later, facilitates the provision of double current ranges in both ammeters and wattmeters.

Essentially the torque produced by the dynamometer is proportional to $I_1 \times I_2$ where I_1 and I_2 are the currents in the fixed and moving coils respectively. This relationship holds for the instantaneous values of these currents so that inherently it is an R.M.S. indicating instrument and in the case of the voltmeters and ammeters, since the current in all the coils is either the same current or a proportion of it, the scale-shape is approximately a square law.

As the device is air-cored the operating fluxes are low, only of the order of 50 gauss so that unless it is very efficiently shielded it will be seriously disturbed by any external field. In the version manufactured by Sangamo Weston the whole mechanism is surrounded by a cylindrical mumetal shield which is carefully heat treated after fabrication to ensure its shielding efficiency. This whole assembly is then mounted into a further cylindrical shield with one closed end and spaced away

Fig. 3.11.

from the mumetal shield by an air space of approximately half an inch. The outer shield is made of a high quality soft iron, also heat treated, and the shielding is completed by a flat end shield, again of mumetal, so that the whole assembly is a closed pot. The considerable care taken in the manufacture of these shield components is very necessary for two reasons; first, to ensure the effectiveness of the magnetic shielding and secondly, to reduce the likelihood of remanence in the shields causing reversal errors when the instrument is used on d.c. Figure 3.12 shows a Sangamo Weston dynamometer and the shields can be clearly seen. The basic constructions of the voltmeter, ammeter and wattmeter are very similar but each has its special features and these will be dealt with under their separate headings as follows.

The dynamometer voltmeter

As shown in fig. 3.10 the field coils and moving coil are all connected in series and built out to the appropriate resistance with a non-inductive series resistance. In order to keep the current consumption low, the field coils are wound with a large number of turns, as is also the moving coil. In the case of the Sangamo Weston version of this instrument the moving coil is wound with aluminium wire to reduce the weight of the moving system and the typical current consumption of a 250 volt voltmeter would be 30 mA. The number of turns on the field coils in this instance is 1500 per coil and the number of turns on each moving coil is 300. The total inductance of this arrangement when all coils are in series is approximately 0·275 henries. The series resistance required is approximately 8000 ohms, and this is made up of minalpha

Fig. 3.12.

wire wound on thin micalex cards. This construction of series resistor gives negligible inductance, low capacitance and high heat dissipation. This form of voltmeter is normally only offered for use over a restricted frequency range since the high inductance of the coils would lead to a significant frequency error. This error can be readily calculated for the above design example as follows:

$$Z \text{ (impedance)} = \sqrt{[(R+r)^2 + \omega^2 L^2]},$$

where R = series resistance, r = total resistance of coils and L = total inductance of coils. At a frequency of 1000 Hz this becomes 8500 ohms compared with a d.c. resistance of 8330 ohms, which gives an error of approximately 2%.

For lower voltage instruments say below 75 volts full scale it is usual to reduce the number of turns on the coils in order to reduce the inductance. Since the inductance of a coil is proportional to the square of the number of turns, halving the number of turns reduces the inductance to a quarter of the previous value. If it is still required to develop the same torque the current consumption is doubled, but the net result is an improvement in the frequency error. For optimum performance a number of different designs are necessary to cover the voltage range from say 1·5 volts to 750 volts and it is not possible to make any one instrument cover efficiently more than a small portion of this range. The exact design of coils for the different ranges is a complex interplay between various factors such as resistance, inductance, temperature coefficient and the sheer practical difficulty of available wire gauges. To illustrate this problem the following table give the range of coils used on the Sangamo Weston 6 in scaled dynamometer voltmeter. This instrument is offered to an accuracy of $\mp\frac{1}{4}\%$ of full scale deflection except on the ranges below 10 volts where the increased current consumption gives rise to self-heating errors which make it impossible to achieve this very high accuracy. These are offered at $\mp\frac{1}{2}\%$ of full scale deflection. The self-heating error is caused by changes in the resistance of the coils due to the heating effect of the high current, the heating of the instrument springs due to the passage of current and also due to heat radiated from the coils to the springs. The two effects of heating, one on the coils and the other on the springs, tend to be self-compensating since the springs get weaker with increase in temperature whilst the increased resistance of the coils lowers the total current. It is not possible to balance these two effects sufficiently to allow the highest accuracy to be maintained.

Mechanism No.	Turns on moving coil	Turns on field coil (each)	Sensitivity (mA)
1	300	1500	30
2	220	900	45
3	177	515	65
4	155	450	75
5	123	325	100
6	91	200	152
7	38	130	282
8	28	57	485

Due to the inductance of the coils the dynamometer voltmeter suffers from frequency errors. These may be compensated to a large extent in the same way as for a moving iron voltmeter by shunting the series resistance with a suitable capacitor. This compensation is normally only effective over a limited frequency range and the best value for the capacitor can only be found by experiment. The compensation of very low range voltmeters which have a high current drain is not normally undertaken since the value of shunt capacitor would be impracticable. As an example for a typical low range voltmeter made by Sangamo Weston having a range of $1\frac{1}{2}$ volts the inductance is 0·00047 H. The current for full scale is 480 mA the series resistance is only 2 ohms so that the shunt capacitor required would be approximately 120 microfarads which is impractical for the paper capacitors normally employed.

Dynamometer ammeters

The construction of the ammeter is very similar to that of the voltmeter already described. The circuit is shown in fig. 3.10 and it will be seen that by a system of links the connection to the field coils and shunts can be changed to give a two to one change in full scale deflection.

In order to ensure an accurate range change it is necessary that the field produced by the two field coils shall be reasonably balanced, so during initial testing the field coils are put into circuit one at a time and their positions relative to the moving coil adjusted until the full scale sensitivity obtained with one coil is within 0·2% of that obtained with the other.

The frequency errors of dynamometer ammeters are worthy of note since they exhibit errors due to a phenomenon which does not show up to any marked extent in the voltmeters and wattmeters. This error is due to the mutual coupling between the fixed and moving coils. Since the position of the moving coil relative to the fixed coils changes with deflection the whole assembly behaves rather like a variable mutual inductance and the final resultant error is shown in fig. 3.13. It will

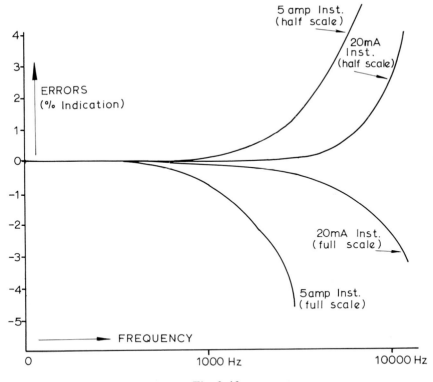

Fig. 3.13.

also be seen that the error is very much range dependent due to the differing turns on the fixed and moving coils. The error also changes sign over the scale as seen from the different curves for full scale and half scale errors, which implies that at some point on the scale the error will be zero. This has been confirmed experimentally and shown to coincide with the point of zero mutual, that is when the moving coil is at right angles to the field coils.

Owing to the variable nature of the frequency error with deflection it is impossible to compensate it completely but considerable improvement can be made by using the circuit shown in fig. 3.14, the value of

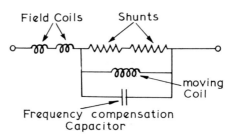

Fig. 3.14.

the required capacitor being approximately determined by the expression $C = L/R^2$ and the actual best value for a particular instrument by experiment.

One version of the dynamometer ammeter which has not so far been mentioned is the single-range milliammeter. This is made by connecting the field coils and moving coil in series. By suitable design of special springs having low resistance and a large heat dissipating area it is possible to make an instrument of this type having a range of 1 ampere full scale. The particular advantage of this construction as opposed to the double range shunted type is that it has a very much smaller frequency error. Tests have been made up to 10 000 Hz and show errors of not more than about 0·25% compared with the d.c. calibration. Attempts have been made to use this type of instrument with a specially designed current transformer, making it multi-range, in an endeavour to exploit the good frequency characteristic. This experiment only met with partial success as the transformer errors exceeded the instrument errors at about 2000 Hz. This is probably due to the rather high V.A. burden imposed by the instrument and also due to the fact that the V.A. burden increases with frequency owing to its largely inductive nature.

This type of instrument has only a limited application but within its frequency range is more accurate and repeatable than the thermocouple type instrument which would normally be the only suitable alternative.

Dynamometer wattmeters

It is in the case of the wattmeter that the dynamometer makes its most valuable contribution. A reliable direct reading wattmeter is a very necessary instrument for measuring the power obtained from or supplied to many electrical devices. The dynamometer is undoubtedly the best instrument for performing this function and when carefully constructed has the additional advantage that its calibration can be checked on d.c. The a.c. to d.c. conversion error of a well-designed wattmeter is zero for all practical purposes.

The construction of the wattmeter is again very similar to that of the voltmeter and ammeter except that the connections to fixed coils and the moving coils are brought out separately. The fixed coils are connected in series with the load and the moving coil built out with a suitable non-inductive resistance is connected across the load. The two alternative methods of connecting this type of single phase wattmeter are shown in figs. 3.15 (*a*) and 3.15 (*b*).

In fig. 3.15 (*a*) where the voltage coil is connected on the load side of the current coil, the current coil also carries the current taken by the

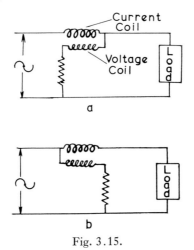

Fig. 3.15.

voltage coil whilst in fig. 3.15 (*b*) the voltage as seen by the voltage coil is larger than that across the load by the voltage drop in the current coil. In most high current range instruments these errors may be neglected but in the special case of low current or low power factor circuits the errors may be considerable. A special version of wattmeter is made which corrects this error and is dealt with under low power factor wattmeters.

The instantaneous torque of a dynamometer, as previously stated, is proportional to the product of the instantaneous currents in the coils and in the case of the wattmeter the deflection depends upon the average product of these currents and therefore upon the mean power. This fact also makes the wattage scale nominally linear.

The wattmeter is liable to suffer from a number of small errors due to phase angle errors in its own circuits which, while they can be made negligible by careful design, are worthy of mention.

If the inductive reactance of the voltage coil is significant compared to the series resistance in that circuit then the current in the voltage coil will lag on the applied voltage by a small angle α so that the power measured instead of being $EI \cos \phi$ will be $EI \cos (\phi-\alpha)$. It can be shown that if the series resistance per millihenry of inductance in the voltage coil is 300 ohms or above this error is reduced to less than 0·1% at a frequency of 50 Hz. This is easily achieved in practice and for use on power frequencies this is normally adequate. It should be noted, however, that at higher frequencies the angle α will increase and the error may well become significant. The angle α also tends to be increased by the temperature compensation network which is usually shunted across the moving coil and when it is known that the instrument is to be used on higher frequencies, say above about 150 Hz, then the temperature compensation circuit is often omitted in the interest of minimum frequency error at the cost of a higher temperature coefficient. For laboratory standard and substandard dynamometers both the frequency and temperature coefficients are normally quoted on an accompanying certificate so that due correction may be made when greatest accuracy is required. Laboratory standards also have a thermometer fitted which indicates the temperature of the mechanism thus enabling the temperature coefficient to be applied to the best possible accuracy.

The low power factor or L.P.F. wattmeter

As previously mentioned both methods of connecting a wattmeter to a load shown in figs. 3.15 (*a*) and 3.15 (*b*) involve inherent errors which are usually negligible where the current range is high but which may become significant where the current or power factor is low. To overcome this problem a special form of wattmeter is made known as an L.P.F. wattmeter. In this device the field coils incorporate a second winding, wound turn for turn with the main winding. This additional winding is connected in series with the voltage coil in such a direction that the field produced is in opposition to the main flux. The circuit of this arrangement is shown in fig. 3.16 and it will be seen that the current taken by the voltage coil, in passing through the compensating winding produces a flux which exactly balances out the error.

It is usual on the L.P.F. wattmeter to provide a switch position on the voltage switch for an ' uncompensated ' circuit in which the additional winding is eliminated. This is for use when the instrument is being checked, under what is often known as ' phantom loading ' condition. This is an arrangement whereby the voltage and current coils are supplied from independent sources such that the voltage coil current does not flow through the current coil circuit and consequently no compensation is required.

L.P.F. wattmeters are normally designed to give full scale indication with rated current and rated volts applied when the power factor is 0·2 or 0·1. This means that the wattmeter will give full scale indication on d.c. or at unity power factor with either one-fifth or one-tenth of the rated current. In order to increase the sensitivity to this extent the torque for full scale has to be somewhat reduced and the field coils are overrun. This leads to larger friction and self-heating errors and such instruments are not offered to the highest accuracy grade and are not

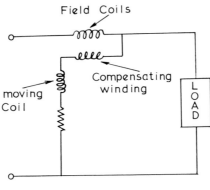

Fig. 3.16.

offered as laboratory standards. In spite of these apparent small disadvantages this type of wattmeter has been found extremely useful for measurements where low power factors are encountered, the slightly lower accuracy grade being more than offset by the improved readability and compensation.

A detailed consideration of the design of a typical low power factor wattmeter is given below:

Ranges: 150 and 300 volts, 2·5 and 5 amperes. Power factor 0·2.

Moving coil: 300 turns, resistance approximately 45 ohms.

Field coils: 72 main turns and 72 compensating turns per field coil.

Resistance of each compensating winding 10 ohms.

The design constants of the particular construction are such that the ampere turns squared required to produce a torque of 100 milligramme centimetres per 100° deflection is 550. Consideration of other factors such as adequate torque to weight dictate that the actual spring torque used shall be 140 milligramme centimetres for 100°.

Therefore
$$(AT)^2 = 550 \times 1 \cdot 4 = 770,$$
$$(AT)^2 = I_{MC} \times N_{MC} \times I_{FC} \times N_{FC},$$
$$I_{MC} = \frac{770 \times 10^3}{300 \times 72 \times 1} = 35 \cdot 6 \text{ mA approx.}$$

Note: The figure of 1 A is the value of I_{FC} used for calculating I_{MC} as the power factor is 0·2.

The instrument is scaled 0–150 watts and the actual watts indicated at full scale on the various ranges is as follows:

Range	Watts (0·2 pF)	Scale multiplying factor
5 A, 150 V	150	1
5 A, 300 V	300	2
2·5 A, 150 V	75	0·5
2·5 A, 300 V	150	1

The potential watts for 150 volt range $= 150 \times 0 \cdot 0356 = 5 \cdot 34$ watts.
The potential watts for 300 volt range $= 300 \times 0 \cdot 0356 = 10 \cdot 68$ watts.
Referring to the circuit diagram shown in fig. 3.16 then using the 5A range only one compensating winding is in circuit the opposing $(AT)^2$ are equal to:

$$I_{MC} N_{MC} I_{CW} N_{CW} = 0 \cdot 0356 \times 300 \times 0 \cdot 0356 \times 72 = 27 \cdot 4 (AT)^2.$$

Using the scaled watts, the $(AT)^2$ per watt is $770/150 = 5 \cdot 133$.

Therefore the opposing watts for the 150 volt and 300 volt ranges $= 27 \cdot 4/5 \cdot 133 = 5 \cdot 34$ watts which exactly counterbalances the potential circuit watt loss. (Note: on the 300 volt range this value of 5·34 watts is multiplied by 2 due to the range multiplying factor.)

On the 2·5A range the two compensating windings are in series and similar expressions can be written showing that the opposing $(AT)^2$ is 54·8.

$\therefore$ The equivalent opposing watts $= 54 \cdot 8/5 \cdot 133 = 10 \cdot 68$ watts.

Self-heating errors

All air-cored dynamometers are likely to suffer from self-heating errors, since to obtain reasonable working fluxes implies a fairly high wattage dissipation in a small space. By careful design these errors can be minimized but may need the addition of some compensation in order to reduce them to negligible proportions. The requirement to achieve this is particularly necessary in the case of wattmeters, since if they are to be used for the checking of kilowatt hour meters they may have to meet the stringent tests laid down under the Electricity Supply Meters Act (E.S.M.A.). This specification requires that a 6 in. scale dynamometer wattmeter shall be left on circuit for a period of half an hour at rated voltage and two-thirds rated current. The additional error (or self-heating) caused by this test must not exceed 0·2% F.S.D. The two main causes of any self-heating error which may occur due to this test are first the warming up of the instrument springs due to the heat radiated by the field coils. This tends to make the instrument read high as the springs weaken slightly as their temperature increases. (The change is of the order of 0·04% per °C.) Not very much can be done to eliminate this error except by design in keeping the wattage dissipation of the field coils as low as possible and also by ensuring that the heat transferred to the springs is kept as low as possible.

The other major part of the self-heating error is caused by the tendency of the materials used for the series resistance of the voltage coil to have a slightly negative coefficient of resistance. This may be compensated for by including a small amount of copper in the series resistance, and so positioning it that it takes up the general temperature of the resistance sheets forming that series resistance. As the resistance sheets warm up due to the wattage dissipated in them, they will decrease slightly in resistance but by the adjustment of the amount of copper included in the circuit the error can be reduced to negligible proportions.

Multi-element wattmeters

For the measurement of power in two or three-phase alternating current circuits the classic method is the 'two wattmeter method' in which either by the use of two wattmeters or by the sequential connection of a single wattmeter the total circuit power is found as either the sum or difference of the two readings. A circuit showing the connections for this method as applied to a three-phase star connected load is shown in fig. 3.17. The method is equally valid whether the load is balanced or unbalanced.

However, where this type of measurement is required frequently, a more convenient form of instrument is available. This is the two-

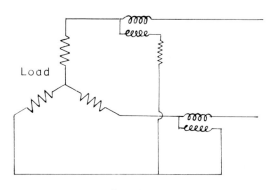

Fig. 3.17.

element wattmeter which consists essentially of two separate single-phase wattmeters but with the moving coils mounted on a common staff, so that the total is obtained automatically with a single pointer reading.

The two wattmeter elements comprising this instrument are completely independent except for the common staff on which the moving coils are mounted and are shielded both magnetically and electrostatically from each other to prevent unwanted interaction.

For the measurement of the power in a three-phase four wire circuit three wattmeters would normally be required but for this particular application a three-element wattmeter is also made and is similar in all respects to the two-element version except for the addition of the third element whose moving coil is also mounted on the common staff.

Most of the remarks concerning design and construction of the single-phase wattmeter apply equally to the two- and three-element versions. In addition, however, the multi-element versions present at least one additional design and manufacturing difficulty, which is to ensure that the two or three elements when energized individually deflect according to a linear scale shape. If this requirement is not met then the sumimation reading obtained when the instrument is measuring three-phase power will not be correct. It is obvious that this requirement calls for extremely close control of design, manufacture and adjustment to obtain the necessary degree of similarity between the elements. As might be expected, however, absolute perfection in this respect is not normally attainable and it is usual for the multi-element wattmeter to be offered to a slightly worse accuracy than its equivalent single-element version. This is not such a disadvantage as might at first appear, since the use of two or three separate wattmeters would involve the possible addition of their separate errors. The net result is usually in favour of the multi-element version which has the additional virtue of convenience in not having to make two or three simultaneous or sequential readings.

Most of the foregoing remarks concerning dynamometers have been concerned with the air-cored versions of the instrument. Much of it is equally true of the iron-cored versions which are made by a number of manufacturers. The primary advantage of the iron-cored dynamometer is the higher efficiency of the magnetic circuits which can be attained due to the presence of the iron and hence the lower watt loss or higher torque which can be achieved. The design of an iron-cored dynamometer needs special care in the design of the magnetic circuit and in the selection of the material to ensure that the various iron losses are minimized and do not result in eddy current and hysteresis errors of such a magnitude as to render the instrument unduly frequency sensitive. Several manufacturers have made good designs which are capable of precision accuracy but the iron-cored dynamometer probably finds its greatest application as a switchboard instrument where the accuracy requirements are not quite so onerous. It can also even be made in a circular scale version and has been applied to aircraft use, particularly as a watt-varmeter for the monitoring of the load sharing on multi-alternator systems.

Special versions of dynamometer

One variation which has been made for special use is the single-phase var-meter in which by means of a series inductance connected as shown in fig. 3.18, the phase of the current in the voltage coil circuit is

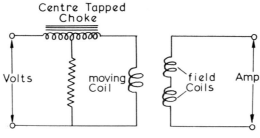

Fig. 3.18.

made to lag by exactly 90° on the applied voltage so that the instrument only responds to the quadrature component of the current in the load circuit. This instrument is extremely sensitive to frequency changes so it is only usable at precisely the frequency at which it was adjusted (usually 50 Hz). It has, however, found limited application in special iron testing circuits.

Special versions have also been made with crossed moving coils for such purposes as the measurement of capacitance, phase angle, etc. but the application of these is so limited that they are not much more than laboratory curiosities.

Electrostatic instruments

The electrostatic instrument makes use of the forces of attraction which occur between oppositely charged plates and the forces of repulsion occurring between similarly charged plates. The normal commercial or switchboard version of this instrument uses both these forces, being in the form of a quadrant electrometer. It consists of four metal quadrants separated by airgaps and arranged to form a metal box in which a metal vane is suspended via normal pivots and jewels or a taut suspension. A diagrammatic view showing the principle of construction is given in fig. 3.19.

Two quadrants are connected to one side of the supply and the metal vane and the remaining quadrants to the other. The torque produced is

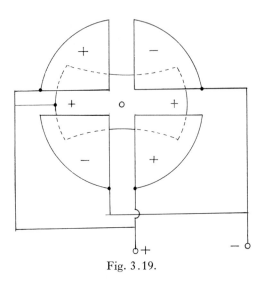

Fig. 3.19.

proportional to the square of the applied p.d. giving a square law scale cramped at the lower end. Where it is required to lower the voltage range the torque can be increased by mounting a number of the quadrant arrangements one above the other with the sectors on a common staff. With multiple construction of this nature instruments are made with a range of about 0–150 volts. Damping is usually pneumatic although both magnetic and oil damping are sometimes employed. The instrument is suitable for use on both a.c. and d.c. and reads true R.M.S. values and is independent of waveform. Its principle advantage is its high input impedance and negligible power consumption. Owing, however, to its rather delicate construction and poor torque to weight ratio when made as a low voltage instrument it is not widely used below voltages in the kilovolt range. Special versions are made for the measurement of voltages up to several hundred kV and in this area of measurement it is probably the best method available.

Induction instruments

These instruments will only function on an a.c. circuit and their primary advantage lies in the fact that a scale angle of some 300° can be achieved fairly readily. Their principle of operation relies on the torque which is produced by the reaction between a flux proportional to the current or voltage to be measured and eddy currents induced in a metal disc or drum which is free to rotate restrained only by a control spring which supplies the restoring force.

Since the deflecting force depends on the reaction between a current and a flux it is necessary to produce a phase difference between them. The method by which this effective phase difference is obtained gives rise to the two general classifications of induction instruments which are: (1) the shaded pole type and (2) the Ferraris type.

1. *Shaded pole type*

A diagrammatic representation of this type is shown in fig. 3.20.

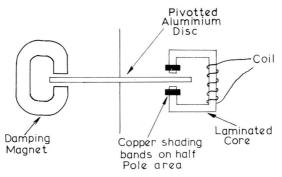

Fig. 3.20.

The addition of a copper 'shading' band to one-half of the pole acts like a short circuited secondary on a transformer and causes the flux produced by the 'shaded' pole to lag behind the flux produced by the unshaded portion. The reaction in the disc can therefore be considered as two torques, one produced by the reaction between the flux provided by the shaded pole and the eddy current provided by the

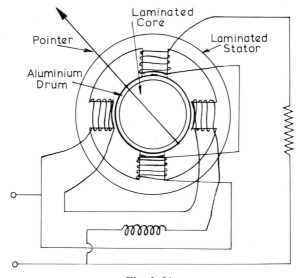

Fig. 3.21.

unshaded pole and the other torque by the reaction between the flux provided by the unshaded pole and the eddy current provided by the shaded pole. Unfortunately these two torques are in opposite directions so that the net torque is only the difference between them.

2. *The Ferraris type*

This instrument works in a manner very similar to that of an induction motor. A rotating field is produced by two coils on laminated polepieces. The necessary phase displacement is caused by including an inductance in series with one coil and a resistance with the other. The rotating field produced induces currents in an aluminium disc or drum causing it to rotate against the restraining force of a spring. A diagrammatic representation of this type of instrument is shown in fig. 3.21.

Induction type instruments are not widely used except where the facility of a scale approaching 360° is important, as the construction is expensive and subject to a number of errors particularly with frequency.

CHAPTER 4
design

1. *Torque equations*

(*a*) *Permanent magnet moving coil*

THE moving coil instrument is the only one of the various instrument types which lends itself to reasonably precise design techniques. The torque equation is derived from the fundamental relationship that the force F on a conductor carrying a current I in a magnetic field of strength H is $HIl/10$ where l is the length of a straight conductor perpendicular to the field.

If the moving coil has N turns and the mean distance of the sides of the coil from the centre of rotation of the coil is r the torque equation becomes:

$$2r \times NHIl/10,$$

which may be rewritten as $NHI/10 \times$ area of coil.

If r and l are in centimetres and I in amperes and H in lines per cm^2 then the torque T is in gramme centimetres. At least one manufacturer uses this equation in a simplified and perhaps more readily remembered form as:

$$T = INK.$$

In this case K is what is known as the flux constant for a particular size or type of mechanism and embodies the coil area and the actual flux produced by the particular size and type of magnet used. The value chosen for K is such that when I is in milliamps then T is in milligramme centimetres. A typical value of K for a miniature instrument mechanism would then be in the region of 0·4 to 0·5.

(*b*) *Moving iron instruments*

No completely satisfactory torque equation has been developed for moving iron instruments largely due to the apparent impossibility of calculating the magnetization which occurs in the fixed and or moving iron when they are placed in a non-uniform field. The equations which follow make certain assumptions which are not completely borne out in practice but will give some guide to the likely performance for early design considerations.

Attraction type

Let us suppose that the arrangement is as shown diagrammatically in fig. 4.1 and that the field H is produced when a current I flows in the

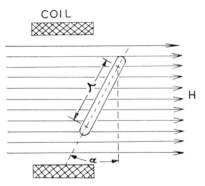

Fig. 4.1.

coil and that it is evenly distributed in a direction parallel to the coil axis. Then if the initial angle between the axis of the soft iron and a direction normal to the field is α the magnetization of the iron is proportional to the component of H which is along its axis, which is $H \cos(90 - (\theta + \alpha))$ or $H \sin(\theta + \alpha)$. The force acting on the disc to pull it into the coil is therefore proportional to $H^2 \sin(\theta + \alpha)$. If it also assumed that the disc permeability is constant then this force is proportional to I^2.

If this force F acts at a distance r and then the torque T is equal to $Fr \cos(\theta + \alpha)$. Therefore the torque T is proportional to

$$H^2 \sin(\theta + \alpha) \cos(\theta + \alpha)$$

which is proportional to $I^2 \sin 2(\theta + \alpha)$.

If the instrument is spring controlled then the spring torque $S = k\theta$ and since a steady deflection will be obtained when $S = T$ then the scale shape for the instrument is given by the relationship:

I is proportional to $\sqrt{\{\theta/\sin 2(\theta + \alpha)\}}$.

Consideration of this relationship shows that the maximum possible scale angle is $90°$ which occurs when α is zero and that the available deflection angle is $(90 - \alpha)$. It is also obvious that the higher the value of α the more linear the scale becomes. The final instrument therefore is a compromise between these two opposing factors, a scale angle of between $60°$ and $80°$ being typical.

This type of moving iron instrument has one particular advantage in that the coil can be made very flat and therefore have a quite low inductance which makes it particularly suitable for voltmeters or for instruments required to operate over a frequency range.

Repulsion type

If for the purposes of this consideration the repulsion instrument is reduced to two parallel rods magnetized by a surrounding coil such that they repel each other and it is assumed that the distance between the rods is small compared with their length then it is reasonable to treat the forces as acting simply between adjacent poles. This layout is shown in fig. 4.2 from which it may be seen that the force

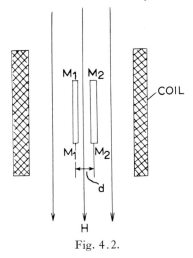

Fig. 4.2.

$F = 2m_1 \times m_2/d^2$, where m_1 and m_2 are the strengths of the poles and d the distance between them. This may be rewritten as

$$2m_1 m_2 \bigg/ \left(2 \sin \frac{\alpha+\theta}{2}\right)^2.$$

Therefore the torque (T) at a distance

$$r = Fr \cos(\theta+\alpha) = \{2m_1 m_2 \cos(\theta+\alpha)\}/4r^2\{\sin(\theta+\alpha)/2\}^2,$$

which, if m_1 and m_2 are proportional to I, is proportional to

$$I^2 \cos(\theta+\alpha)/\{\sin(\theta+\alpha)/2\}^2.$$

for a spring controlled instrument the control torque $S = k\theta$ and for steady deflection $k\theta = T$. Therefore θ is proportional to

$$I^2 \cos(\theta+\alpha)/\{\sin(\theta+\alpha)/2\}^2.$$

or I is proportional to

$$\sin\frac{(\theta+\alpha)}{2} \sqrt{\{\theta/\cos(\theta+\alpha)\}}.$$

In this instance of course α cannot be zero as the two irons cannot be coincident and it should be noted that the maximum scale angle is not now 90° as in the case of the attraction instrument but $(180-\alpha)$, α being typically 20 or 30°.

(c) Dynamometers

With the very short coils that are normally used for commercial electrodynamic instruments the calculation of their constants is not usually practical but a rough approximation can be made involving the use of the mutual coupling M between the fixed and moving coils. It may be shown that when a coil is in the field produced by a second coil its potential $V = \Phi NI$ where N is the number of turns in the coil, I the current and Φ the flux passing through it. If the coil is moved so as to alter the flux then $dV = F\,ds$ where F is the mechanical force and ds the element of distance moved in the direction of the force.

This may be rewritten as $F = dV/ds$ and similarly $T = dV/d\theta$. If I is constant then substituting for V we get

$$T = NI(d\Phi/d\theta).$$

It can also be shown that the linkages $N\Phi$ in the circuit are equal to M (the mutual inductance) times the current in the other coil, so that

$$T = I_1 I_2 (dM/d\theta).$$

If I is in amperes, M in henries and θ in radians then T (in gramme cm) $= 10\,200\ I_1 I_2 (dM/d\theta)$. Using this equation therefore if the rate of change of mutual inductance can be calculated then the torque of the system can be derived.

Various authors have published both calculations and experiments using this formula and while the agreement between calculation and experiment is not perfect a working design can be established which will form the basis for experimental improvement.

(d) Electrostatic instruments

The torque of an electrostatic instrument may be derived from the well-known relationship that the charge Q on a capacitance C charged to a voltage V is given by $Q = CV$.

The energy required to charge the capacitor is $\int_0^v V\,dQ$ which by substitution

$$\frac{1}{C} = \int_0^Q Q\,dQ = \frac{1}{C}\frac{Q^2}{2} = \frac{1}{2}CV^2.$$

Now if the charge remains constant but the distance between the plates is changed then the change of energy $T\,d\theta = d(CV^2)/2$ therefore

$$T = (d/d\theta)(CV^2/2) \text{ but } Q = CV$$

therefore

$$T = \frac{Q^2}{2}\cdot\frac{d}{d\theta}\cdot\frac{1}{C} = \frac{Q^2}{2C^2}\frac{dC}{d\theta} = \frac{V^2}{2}\frac{dC}{d\theta}.$$

To put units to this equation, T is in dyne cm when V and C are in electrostatic units; therefore allowing the appropriate conversion factors

$$T \text{ (in gramme cm)} = 0{\cdot}0051 \ V^2 \ (dC/d\theta),$$

when C is in microfarads and V is in volts.

Here again this torque equation has been verified by comparison between theoretical and measured values and the agreement, whilst not perfect, is sufficiently good to allow a tentative design to be established.

Equations of motion

One of the prime requirements in the design of electrical measuring instruments is that, having made sure of adequate mechanical torque or, better, torque to weight ratio, the instrument should reach its final steady indication quickly and without undue overswings.

The basic design techniques necessary to achieve this end are common to all instrument types although due account must be taken of the manner in which the developed torque varies with deflection. The simplest case is the permanent magnet moving coil instrument in which, if we assume a linear scale, the torque is developed immediately current is applied and remains constant over the total deflection.

If we consider a moving coil instrument which is unenergized and at rest, if a torque is suddenly applied then during the time that the pointer is in motion this torque is made up of three components:

(a) The torque required to accelerate the movement from rest. This is $M(d^2\theta/dt^2)$ where M is the moment of inertia of the moving system and θ is the deflection at any time t after the torque has been applied.

(b) The torque required to overcome the restraint due to damping. This can be expressed as $D(d\theta/dt)$ where D is what is known as the damping coefficient.

(c) The torque required to overcome the torque of the spring. This is proportional to the angular deflection and therefore can be represented by $S\theta$, where S is a constant.

Thus the equation of motion becomes:

$$T \text{ (the applied torque)} = M \frac{d^2\theta}{dt^2} + D \frac{d\theta}{dt} + S\theta.$$

The solution of this equation will enable the relationship between time and deflection to be established. A solution can be obtained by the use of operational calculus and the roots of the equation are:

$$p = -D/2M \pm \sqrt{\{(D/2M)^2 - S/M\}}.$$

This equation becomes easier to manipulate if we make two further substitutions:

An undamped moving system has a natural period of
$$T = 2\pi\sqrt{(M/S)} \quad \text{or} \quad \sqrt{(S/M)} = 2\pi/T.$$
If we also give the symbol n to the ratio between the actual damping coefficient D and the damping coefficient for critical damping then the roots of the equation can now be rewritten as equal to
$$2\pi\{n \pm \sqrt{(n^2-1)}\}/T.$$
It can now be seen by consideration of this equation that when n is equal to 1 the damping is critical whilst if it is less than 1 the roots are complex which signifies an oscillatory condition. If n is greater then 1 then the instrument is overdamped.

The numerical solution of these equations for the three possible conditions of damping, i.e. underdamped, critically damped and overdamped, is too complicated to show here but graphs illustrating deflection against time for the three conditions are shown in figs. 4.3, 4.4 and 4.5.

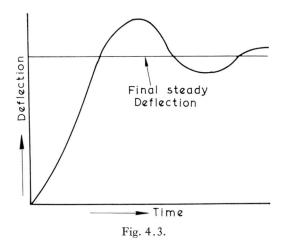

Fig. 4.3.

From these further curves can be derived such as are shown in fig. 4.6 where the time to reach a certain deflection is shown in terms of the fraction of the natural period versus the damping factor n.

This information is very useful in determining what is known as 'response time'. It should be noted that according to the original equation of motion, theoretically an instrument will not come to rest until an infinite time has passed but in fact due to friction between pivots and jewels the instrument does settle in its final position relatively quickly and the response time is usually taken to mean that time which elapses after the application of a step input before the pointer gets within and stays within its declared accuracy of its final reading. It

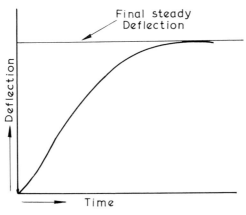

Fig. 4.4.

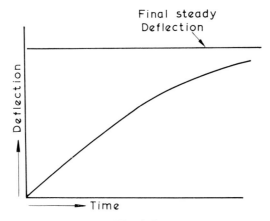

Fig. 4.5.

should also be noted that this definition implies that the same instrument offered to different accuracies can have differing response times, greater accuracies resulting in slightly longer response times.

It is usual to make general purpose instruments slightly underdamped since this gives the shortest response time, particularly if the first overswing does not take the instrument past its final indication by more than its accuracy limit. This condition whilst ideal is difficult to control economically and most commercial instruments will be found to have more overswing than this. A figure of 5 to 10% for the first overswing is often used as a design target so that small variations from these values will not cause some instruments to be overdamped.

It must be remembered that the foregoing comments apply only to instruments having linear current/deflection characteristics employing

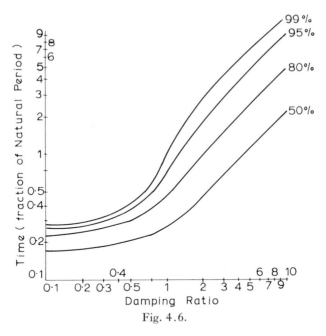

Fig. 4.6.

electromagnetic damping. The problems of non-linear parameters such as are encountered with, for example, shaped polepiece instruments is beyond the scope of this volume, involving various approximations and empirical data peculiar to a particular manufacturer's construction. Similarly, other forms of damping such as air damping and fluid damping whilst partially amenable to calculation, when associated with non-linear torques such as are usually encountered in moving iron and electrodynamometer instruments, become so complex that it is usual to employ empirical data derived from previous experience.

Special purpose apparatus has been developed for the measurement of instrument response times and is frequenctly used at the prototype stage to check the performance of a model against the calculations.

The most commonly employed principle, which has been made considerably easier by the advances in electronics, is to view the instrument by means of a very short pulse of light produced a known and predetermined time after the application of a step input to the instrument. This has the effect of 'freezing' the motion of the instrument at an instant in time. By repeating this procedure with varying times the whole motion of the instrument from the initiation of the input pulse until it comes to rest may be explored. The process may be repeated after a small modification has been made to the instrument so that the precise effect of the modification can be determined and used where necessary to predict the degree of further modification required to obtain a particular performance.

Bearing systems

In the foregoing comments on damping and response times the effect of friction in the instrument bearing system has been ignored, which is generally quite justified since friction torques are normally kept to a very small percentage of the full scale deflection torque. However, friction torque in pivot and jewel bearing systems is the most important single factor in determining the lower limit of sensitivity which can be achieved with any particular design. Until a few years ago most commercially available instruments employed a bearing system consisting of a hardened steel pivot with a conical end running in the bowl of a glass or sapphire jewel. Different manufacturers have their own particular detail variants of this system but as a very approximate figure the friction error for instruments with weights of moving parts up to about 0·5 gramme is equal to kW where $k \simeq 0·15$. This gives the friction torque in milligramme centimetres where W is the weight in grammes.

This is true when the axis of rotation is vertical and all the weight is resting on the bottom jewel. If the axis is horizontal, as when the instrument is mounted on a vertical panel, this figure must be increased by a factor of approximately 4. Consideration of these factors together with the design parameters of a typical miniature instrument show that, if the friction error is to be kept below the point where it becomes objectionable and liable to interfere significantly with the accuracy of reading, the full scale spring torque must be not less than about 10 milligramme centimetres.

Further consideration of this minimum torque shows that this limits the minimum full scale sensitivity to around 25 to 50 μA.

It must be appreciated that these figures are very approximate and related to a typical design of instrument and may be improved on by a factor of four or five times if very high resistance moving coils can be tolerated and other techniques involving extra large magnets etc. are not ruled out by cost.

This limitation of instrument sensitivity due to bearing friction is one of the principal reasons why in the past few years considerable interest has been shown in the application of taut suspensions to commercial instruments.

Taut band suspensions

The modern taut band suspension is illustrated diagrammatically in fig. 4.7. It will be seen that the conventional pivots, jewels and springs have been replaced by two metal ribbons, one at each end of the moving coil, which are maintained in tension by two cantilever springs. The metal ribbons or taut band provide bearings, restoring torque and electrical connection. The system is by no means of recent origin having been applied to more specialized mechanisms such as the

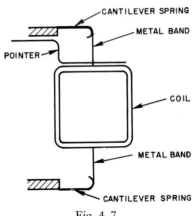

Fig. 4.7.

vibration galvanometer for many years. But it is only as a result of improved materials together with the need for more sensitive and still robust instruments that it has found recent application to the more common commercial instruments.

The use of a taut suspension eliminates the friction errors associated with pivots and jewels, leaving only the molecular friction within the band which occurs as it is twisted. By careful design this friction can be reduced to completely negligible proportions, thus enabling considerably more sensitive instruments to be made than is possible with the pivoted design.

The suspension ribbons can be made of a number of materials but the most commonly employed are beryllium–copper and a specially developed nickel platinum alloy. This latter alloy has the added advantage of extremely good corrosion resistance, which is very desirable in view of the very small cross section of the suspension, which is commonly only two or three thousandths of an inch wide by a few ten thousandths of an inch thick. Ribbons of this size are of course not particularly robust and could easily be broken by careless handling of the instrument if not protected. Most manufacturers of taut band instruments employ some form of shock protection for the delicate suspension and some very ingenious designs are made but they all rely on a similar principle of restricting the movement of the coil both vertically and laterally so that the suspension cannot be overstressed. Since the breaking strain may be only of the order of 100 gramme, with a working stress of at least half this value the protection involves precision parts with only small working clearances. The principle of the suspension protection is shown in fig. 4.8. although actual detail design varies considerably between manufacturers.

When a taut band instrument is subjected to a shock of, say, $100g$ what actually occurs to the suspension is fairly complex, but the high

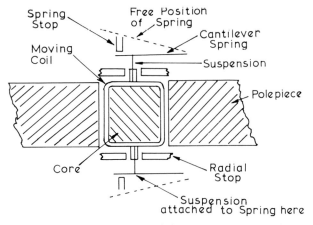

Fig. 4.8.

speed camera has been used to study the phenomenon and has shown, for example, in the simple case where the shock is applied in line with the suspension that fracture of the suspension usually occurs as the upper cantilever spring snaps back in an upward direction. By careful design, including reducing the inertia of the cantilever spring to a minimum, this breakage can be prevented and at least one manufacturer has made instruments which will withstand shocks of several hundred g without suspension fracture.

The provision of a virtually frictionless bearing system as found in the taut band does not mean that instrument sensitivity can be increased indefinitely. As previously indicated there is a lower limit to the size of ribbon which can be employed, owing partly to handling problems and partly to the need to maintain a minimum working tension in the suspension. If the tension is reduced below a certain level the instrument is likely to show position and repeatability errors which are associated with the 'sag' which occurs in the suspension due to the weight of the moving element when the instrument is put in the position where the suspension is horizontal.

This 'sag' is given by the expression:

$$\text{Sag } (S) = \frac{Wl}{2t},$$

where W = weight, l = length of 1 suspension, t = tension. This sag (measured at the mid-point) is usually limited to not more than one or two thousandths of an inch and therefore puts a lower limit to the size of ribbon which can be employed for a particular mechanism.

Another point which has to be borne in mind is that, as the full scale torque is reduced, balancing the instrument also becomes progressively more difficult. For example, even when the full scale torque (for say 100° deflection) is 10 milligramme centimetres an unbalance of only 0·05 milligramme centimetre can give rise to an error of $\frac{1}{2}$% of full scale deflection.

Very low torque instruments may also tend to be slow in response and overdamped particularly where most of the damping is due to the external circuit resistance, as is often the case with microammeters.

Notwithstanding these slight limitations perfectly satisfactory instruments can be made down to a full scale sensitivity of 10 μA or less and which will withstand the normal handling of an industrial environment. The taut suspension instrument therefore represents a very considerable step forward compared with its pivoted predecessor. It has, however, one particular drawback which must not be overlooked. Although it can be made to withstand shock and vibration probably better than its pivoted counterpart it is liable to show greater errors of indication whilst the vibration is actually present. This is due largely to the extra freedom of movement of the suspended mechanism compared to the pivoted version and normally results in more obvious resonances and consequent errors. Generally, therefore, taut suspension mechanisms are considered unsuitable for use in aircraft or other environments, where continuous vibration is present when readings are required to be taken, although they are likely to suffer less damage from the vibration than an equivalent pivoted instrument.

Temperature errors and their compensation in moving coil instruments

Various parts of a moving coil instrument exhibit changes with temperature and may in certain circumstances result in a considerable error if not eliminated by some form of temperature compensation.

As a straight microammeter or milliammeter (i.e. with only the moving coil in circuit) the modern instrument has quite a good temperature coefficient. The only two components which contribute significantly to any temperature error are the springs with a coefficient of approximately -0.04% per °C and the magnet with a coefficient of approximately -0.02% per °C. Fortunately these errors are in opposite directions as both the magnet and springs get weaker with increase in temperature. The net result is a coefficient of about 0·02% per °C for the instrument which for most practical purposes can be ignored. Where a higher degree of precision is required as for instance in a laboratory standard a form of resistive compensation can be employed as shown in fig. 4.9. By careful calculation and adjustment it is possible to reduce the instrument coefficient to less than 0·005% per °C over a temperature range of say 0°C to 40°C. This degree of compensation is rather expensive to achieve since due to minor variations from instrument to instrument it is necessary to check and adjust the compensation for each instrument individually, which could not be justified for anything but the highest precision instruments.

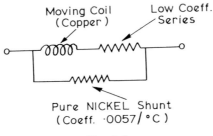

Fig. 4.9.

For shunted milliammeters and ammeters the problems are somewhat different since in an uncompensated instrument the primary temperature error is due to the change in resistance of the moving coil, which if it is made of copper increases its resistance by approximately 0·4% per °C. If the shunt is made of a material such as minalpha, having a negligible temperature coefficient of resistance, the instrument will have an overall coefficient approaching that of copper and becoming less sensitive as the temperature is increased. The usual method of reducing this coefficient to acceptable proportions is to include in series with the moving coil a swamping resistance of some five to ten times the value of the moving coil and made of a resistance material having virtually zero coefficient. The overall resistance of the moving coil plus swamping resistor then has a coefficient of only one-sixth to one-eleventh of the copper by itself and the instrument coefficient is correspondingly reduced.

Voltmeters do not normally present any great problem since the series resistance is usually many times the resistance of the moving coil and providing it is made of a material of negligible coefficient any change in the resistance of the moving coil is almost completely swamped.

The thermocouple millivoltmeter used for temperature measurement is an interesting example of the use of temperature compensation. The circuit of the instrument plus external thermocouple is shown in fig. 4.10. In this circuit there are two primary temperature errors which must be compensated:

First a resistance change due to the moving coil which would have a large effect on the current flowing if left uncompensated. The coil is usually wound with aluminium wire which has a temperature coefficient

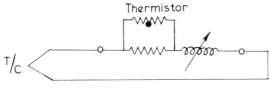

Fig. 4.10.

similar to copper, namely about 0·4% per °C. This is compensated by the use of a 'thermistor' which is made of a semiconducting material having a large negative coefficient of resistance. Unfortuntely this coefficient is rather non-linear but by a suitable choice of series and shunt combination the overall resistance can be made sensibly constant over a reasonable temperature range, say from $-40°C$ to $+70°C$.

The second temperature error is due to the fundamental relationship in all thermocouple circuits that the total e.m.f. in the circuit is related to the difference in temperature between the hot and cold junctions. Since the cold junction is in the indicator any variation in temperature of the indicator will cause a change in the indication unless some form of compensation is included.

It is interesting to note that this particular error is an example of the type which gives rise to an error independent of the indication. Most temperature effects in indicating instruments give rise to errors which are a percentage of indication and therefore can be compensated for by special circuit arrangements. The 'constant' error, however, can only be dealt with by a dynamic component and not a static one.

The component employed for the compensation of the 'cold end' error is a bimetallic spiral, to which the outer end of one of the instrument control springs is attached.

By suitable choice of materials and control of the torque ratio between the two instrument springs the bimetallic coil is arranged to function like a temperature sensitive zero adjuster such that when the instrument is not connected to an external thermocouple it will always indicate the ambient temperature being experienced by the instrument. Simple consideration of this arrangement will show that with an external thermocouple connected, the indication will always be the temperature of the hot junction independent of the temperature of the instrument. An essential feature which must be part of the design of this compensation is to ensure that the temperature of the cold junction and the bimetal compensator are the same, and it is usual to arrange that their positions in the instrument are physically as close to each other as possible. Failure to do this may result in large transient errors during changes of ambient temperature.

Torque to weight ratio

No chapter on instrument design would be complete without some mention of 'goodness factor' or 'figure of merit'. This is some figure which can be used to judge the likely performance of a design before it has been made and also to compare one design with another. Different authorities have used slightly different means of arriving at the figure of torque to weight ratio some insisting that it should be ratio of torque/(weight)^n where n is some power varying accordingly to the type and state of the bearings and other factors between slightly less than 1 to perhaps $1\frac{1}{2}$.

General experience indicates that providing similar types of instruments are being compared then the simple ratio of torque to weight is adequate. Typically the torque for full scale deflection in milligramme centimetres is expressed as a ratio of the weight of the moving parts in grammes. For a good quality modern miniature panel instrument this figure would lie between 30 and 100, a figure less than 20 to 25 being considered likely to cause significant pivot friction errors. The figure, however, is obviously somewhat arbitrary and no exact dividing line can be drawn. It should also be remembered that this 'figure of merit' takes no account of the suitability of the instrument in other respects such as damping, temperature, frequency, etc., so that even a design with a very high torque to weight ratio may be totally unsuitable in other respects. The position of use also has a marked effect on the desirable figure. If the instrument is for use on a vertical panel, i.e. with its pivoting axis horizontal then a minimum of about 50 is generally reckoned necessary.

CHAPTER 5
materials and finishes

THE early experimenters and manufacturers of electrical measuring instruments were rather restricted in the choice of materials available to them, apart from the common metals, brass, iron, etc. their insulating and constructional materials being largely limited to those which were of natural origin such as mica, ceramics, asbestos, cotton, silk, shellac, wood, oils and resins.

Today's engineer is confronted with a bewildering array of synthetic materials as well as a very diverse range of metal alloys which have been specially developed for particular properties. Most of the natural materials are also still in use, either singly or in conjunction with synthetics.

The selection of the appropriate material for any particular function has, therefore, become a complex problem which no engineer is capable of solving unaided in every instance. This situation has resulted in many companies in the setting up of groups or individuals as experts in particular areas, such as metallurgy, chemistry and plastics. It is beyond the scope of this volume to attempt to argue or advise on the merits of particular materials for particular functions, but examples of how the selection and control of materials can be achieved will be given. It is convenient to consider the materials in groups related to their general function in measuring instruments, and in most instances, the comments are equally applicable, whether the instrument is moving coil, moving iron or in fact any of the types previously described.

All the instruments in common use with the exception of the electrostatic type employ some form of coil winding, so it is perhaps appropriate to consider first the requirements for:

Wire

The most commonly used material for all forms of instrument field and moving coils is copper. It is normally used in its pure form (about 99·9%) and this gives a flexible conductor second only to silver in conductivity. It is used in sizes down to about 52 S.W.G. (0·0008 in. diameter) usually with one of the many types of enamel insulation but sometimes covered with cotton, silk or one of the synthetic fibres. For special applications glass and ceramic covered wires have also been used, but these are exceptional and only employed when especially high temperatures are involved.

Until the past few years most enamelled copper wire used the oleo-resinous enamels which have proved to be flexible and adequate insulators with good adhesion to the copper when correctly applied. More recently enamels have been developed based on acetal and similar resins which have largely tended to supersede the oleo type. Both of these classes of enamel need to be removed from the copper before a satisfactory soldered or clamped joint can be made. This operation involving the mechanical removal by machining or scraping is tedious and especially difficult in the case of the finer wires. This has led to the comparatively recent development of so called solderable enamels.

These enamels are based on urethane and their characteristic is such that up to about 200°C they behave very much like the previous enamels, but at this temperature they melt and may therefore be soft soldered without previous preparation. The composition of the enamel is such that at this temperature the melted enamel also forms an effective flux, thus facilitating the necessary tinning of the copper. These solderable enamels tend to have the slight disadvantage that they are not quite so resistant to the effects of solvents as the non-solderable types but the slight extra care needed in handling and in the selection of a compatible impregnating varnish is more than offset by the greater ease with which good soldered joints can be achieved.

A further interesting development of this type of covering is the self-bonding variety in which a further thin covering of a suitable thermoplastic material over the enamel enables a wound coil to be bonded together by the application of heat for a short period. An alternative method of achieving the same result is to pass the wire during winding through absorbent pads which are kept moistened with a suitable solvent. This has the effect of softening the outer plastic covering and bonding takes place as the coil is wound. The evaporation of the solvent completes the process and the pressure between adjacent turns during winding is sufficient to cause the softened coating to bond between turns, the space factor obtained being only slightly worse than that for conventional enamels.

The use of enamelled wires is particularly important for moving coils where the absence of fluffiness is essential and where space factor and weight are at a premium. For many field coils, however, while these factors have still to be considered they are often not quite so critical and here cotton, rayon or silk covered wires are frequently employed, the coils either being varnished as winding proceeds or impregnated on completion, the covering, being absorbent, forming a good carrier for the varnish. This type of construction yields a good firm coil of high mechanical stability which can be used with only a comparatively flimsy former or even with no former at all. It is also a more suitable construction where the voltage between turns is high as the thicker covering when impregnated with varnish forms a good insulation.

Coils for use on high current ranges, such as on moving iron instruments, are often wound with square section wire, and for such uses where the turns are spaced apart, non-insulated copper is commonly used, the material being sufficiently large in section to be completely self-supporting and requiring no former or insulation, being supported only by the end connections.

The other winding wire in common use, although by no means as extensively as copper, is aluminium. Its main use is in dynamometer moving coils and permanent magnet moving coils for certain millivoltmeters where the torques available are low and consequently weight is at a premium. Aluminium has a conductivity of about 60% of that of copper, but the specific gravity is only 2·7 against 8·9 for copper, so that its weight compared to copper for an electrically equivalent coil is only about a half. On the face of it, therefore, it would seem surprising that aluminium is not more widely used in instruments. Unfortunately, everything is not as simple as it sounds; first, although an aluminium coil would be lighter than its copper equivalent, its volume would be about 66% greater and since in instrument design space is also at a premium due to the necessity to work with small air gaps, this factor alone often rules it out. In addition it has two further major disadvantages; it is mechanically much weaker than copper, therefore very difficult to handle in the smaller sizes, and it is also, particularly in the smaller sizes, difficult for making joints.

It is therefore, only used where a detailed consideration of the design shows that its weight advantage is more than enough to offset the volume and handling difficulties. These considerations normally limit its use to low torque moving coil millivoltmeters and dynamometer moving coils.

The range of coverings available on aluminium wire is similar to that on copper, except for the solderable enamels which cannot be used successfully on aluminium. Its most common insulation, however, does not involve the use of an added layer, but is derived directly from the aluminium by anodizing. This is an electrolytic process similar to plating, by which a layer of aluminium oxide is formed on the surface. Aluminium oxide is an excellent insulator and for most instrument purposes, a layer 0·0001 in. thick is adequate. Any attempt to obtain high insulation by thickening this layer unduly on fine wires, leads to cracking during winding as the aluminium oxide is very hard and brittle.

Making joints to aluminium wire presents considerable difficulty owing to the fact that a freshly cleaned surface very rapidly produces a thin transparent film of oxide which defies tinning. Many special solders and fluxes have been produced to overcome this problem; some of them are excellent, but unfortunately involve the use of highly corrosive chemicals. This makes their application to delicate moving coils difficult, if not impossible, since the handling and vigorous treatments necessary to remove any corrosive residue (which if left on a fine wire) could well eat right through and cause the coil to be an open circuit. The final solution, therefore, is often a compromise using as active a flux as can be readily removed by immersion in solvents or ultrasonic cleaning. The joints so made require considerable skill in order to ensure that they will not fail in subsequent use. Partly due to this problem, and partly due to its low tensile strength, aluminium is limited to diameters not less than about 0·004 in. for mass produced instruments, although anodized wire down to less than 0·001 in. diameter has been used for very special applications where cost is of little importance.

For use in the manufacture of moving coils any winding wire must be free from magnetic impurities since the presence of any magnetic material can show up as unexpected changes of deflecting torque or scale shape. Both these effects are accentuated by high gap flux, since the attracting force is proportional to the square of the flux, or by non-uniform fields since the tendency is for the impurity to be attracted towards the stronger part of the field.

The inclusion of magnetic impurities during the manufacture of wire from either copper or aluminium largely occurs during processing, as with modern refining methods the raw material can be obtained almost completely iron free. The likely source of impurity is usually the steel dies through which the wire is drawn. Fortunately, most of the impurity arising from this source can be removed by pickling in suitable acids which selectively remove the unwanted iron. It is standard practice to purchase wire with a stated maximum magnetic susceptibility but equal care is needed during the manufacture of the coil to see that further magnetic impurity is not introduced.

As suggested before, this problem is most likely to occur in permanent magnet moving coil instruments particularly where the gap flux is unusually high or non-linear. For such uses it is normal to specify a magnetic impurity check on the finished moving coil. A piece of apparatus which has been used for such measurements is shown in fig. 5.1. It consists of a solenoid with an iron core having a hemispherical pole brought out at one end. The solenoid is energized from a d.c. source. In the chimney-like structure, which also provides protection from draughts, provision is made to suspend the item under test, such as a moving coil, from a fine suspension. The test is zeroed before the solenoid is energized by setting the suspension to a zero line marked on the side of the ' chimney '. When the solenoid is energized, if the specimen contains any magnetic impurity, it will be attracted towards the centre pole and the level of impurity can be read off against the calibration. The actual effect in practice of any particular impurity level can only be determined empirically, but a rather extreme example, which actually occurred, may serve to illustrate the problem.

An extremely sensitive low resistance moving coil mechanism was

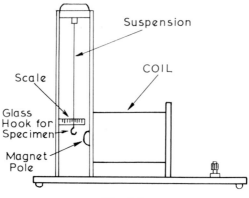

Fig. 5.1.

required for use in a special detector and it was considered that a 1½ in. diameter mechanism normally used for moving coil relays would offer the best chance of achieving the desired sensitivity. The instrument was fitted with control springs having a torque of 20 milligramme centimetres per 100°.

When the first model was checked it was found that the current required to deflect the mechanism was about twice that which had been calculated. The only known phenomenon which could account for the discrepancy was magnetic impurities in the moving coil assembly, and this was proved to be the case by ageing the magnet by some 10 to 20% and remeasuring the sensitivity. The sensitivity had actually improved. The improvement was due to the fact that whilst the deflecting torque was reduced in direct proportion to the reduction in gap flux, the parasitic torque due to magnetic impurity was reduced in proportion to the square of the reduction in gap flux and the net result was a gain in overall sensitivity. By more careful selection of material and testing of finished moving coils, subsequent models were made which did not exhibit this phenomenon to any noticeable extent.

Another aspect of wire which needs consideration, more especially in the finer gauges, is the quality of insulation. For use on permanent magnet moving coils, as previously stated, space is always at a premium and it is usual to specify wire with the minimum practical thickness of insulation. This increases the tendency of the enamel to what is known as 'pinholing' which are small areas along the wire where the enamel is either very thin or sometimes non-existent. These areas occur due to defects during the enamelling process, largely caused by tarnish or dirt on the surface of the bare wire or by the inclusion of foreign bodies such as dust or free solids in the enamel. The insulation requirements between turns on a moving coil are not particularly onerous, but if a high number of pinholes are present in the wire, the chances of shorting turns are obviously increased, and it is therefore necessary to specify some minimum standard which has been found both acceptable in use and practical to achieve during the wire manufacture.

Further details of these tests and the required quality can be found in the relevant British Standards for copper winding wires B.S. 156 for oleo-resinous enamels and B.S. 1844 for the acetal resin enamels. Basically the pinhole test consists of passing the wire through a mercury bath with a potential of 50 volts connected between the mercury and the wire. Any pinholes will allow a current to pass, and this is indicated and counted. The standard is set as so many pinholes per yard when tested in the prescribed manner, and it has been found in practice that wire meeting this minimum requirement gives little trouble with shorting turns on finished coils. For special applications where the conditions of use are more exacting, some wire manufacturers are prepared to guarantee a higher standard than the minimum laid down in the British Standard, usually at a slightly enhanced price to cover the extra care required in handling.

In addition to the factors already mentioned other mechanical parameters such as minimum bending radius, tensile strength, etc., all of which are covered in the appropriate British Standard, may need to be considered. To sum up, in the specification of wire for any particular application some or all of the following points will need attention, and to ensure a consistently satisfactory end product it may be necessary to provide facilities for checking many of these parameters on a regular quality control basis.

(1) Type of material—copper, aluminium, etc.
(2) Type of covering—oleo resin, cotton, silk, etc.
(3) Diameter of bare wire with limits.
(4) Diameter of wire over covering with limits.
(5) Resistance in ohms per yard with limits.
(6) Permitted magnetic impurity level.
(7) Maximum number of pinholes per yard.
(8) Mechanical features such as tensile strength etc.

Wherever possible purchase of wire to the limits as laid down in the appropriate British Standard is recommended as they represent good commercial limits. However, certain special requirements sometimes demand material having characteristics better than, or not covered by, the British Standard. In this case negotiation with the supplier is the only course of action.

Metals

The majority of metals used in the construction of instruments do not present any particular problems in their specification and control, and for most uses the materials are adequately covered by British Standards. It would be quite impossible to mention all the metals which are used in instruments by different manufacturers, but some of the more common ones are listed below with comments on special features and reference to the relevant British Standard where one exists.

Mild steel

This is used mainly for magnetic circuits and for this purpose should be of low carbon content, since it is required to be magnetically soft. It has been found that the addition of up to 2% lead improves the machinability of the material without detracting from its magnetic performance. Where the use is for shielding purposes Swedish iron is simetimes specified. This is a specially pure grade having excellent magnetic properties.

Electrical sheet steels

There are a considerable number of sheet steels used in the form of laminations in instruments and instrument transformers. For general use silicon iron alloys are most common and are marketed under various trade names such as ' Stalloy ', etc.

Where the demand is for maximum initial permeability or minimum losses on a.c. the nickel iron alloys such as ' mumetal ', ' Permalloy ', etc. are frequently specified. In general the full properties of the second group are only obtained when the parts are correctly heat treated after all machining or fabrication has been completed. The selection of an appropriate material is best undertaken by careful consideration of the requirements such as initial permeability, losses, saturation value, cost, etc. and then consulting manufacturers' catalogues to determine which of the many alloys most nearly meets the requirements. The quantity of material used by instrument manufacturers does not generally justify the manufacture of a special alloy so that the choice is restricted to those which are already available.

Brasses

For most instrument purposes the normal commercial range of 60/40 brasses as covered by B.S. 249 is adequate. They are used mainly for their non-magnetic properties, their easy machinability and ability to take a high finish. The strength is adequate for most screws, nuts, pillars, washers, etc., and in general for industrial instruments not subjected to moist or corrosive atmospheres they can be used unprotected, although a very thin nickel or other plating is sometimes used purely for appearance. Where greater strength or improved corrosion resistance is required the so-called naval brasses which consists of a 60/40 copper–zinc brass with about 7% of the copper replaced by tin are used. There are many other brass alloys also containing lead, tin, etc. but as a general guide to the simple copper–zinc brasses the hardness and strength increases and the ductility decreases with the zinc content.

Die-casting alloys

A number of different die-casting alloys are sometimes used in the manufacture of instrument parts. They range from soft tin-based alloys up to the brasses, but by far the most commonly used are the aluminium based and zinc based alloys.

Where the quantities required justify the tooling, die-casting techniques provide an economic method of producing complex shapes and a high degree of dimensional repeatability can be achieved. Various aluminium alloys are used, one of the most popular being a 13% silicon aluminium as covered by B.S. 1490. This alloy casts well and can be readily anodized where good anti-corrosive properties are required. Its strength is adequate for most instrument structures and it machines fairly readily where extra close tolerances are required. A point to be remembered about aluminium alloys is that like pure aluminium they rapidly form an oxide coating on the surface, and if they are included in an electrical circuit special precautions will be needed to ensure an adequate contact.

The other popular range of alloys are the zinc based mazaks. These are covered by B.S. 1004. For a long time after its introduction mazak had the reputation of being unreliable and likely to crack, warp or corrode. This was certainly true in many instances, and it was not unknown for mazak components subjected to moist atmospheres to crumble into a heap of grey powder. Intensive research carried out some years ago established that this type of failure was entirely due to small changes in composition and the presence of small traces of impurity such as tin which have a disastrous effect on the properties. When considering the possible use of mazak, therefore, it is well to remember these problems and to ensure that adequate testing facilities are available such as a spectrograph, because the impurity level likely to cause trouble is not readily detected by any other means. For example, the maximum tin percentage which can be tolerated is only about 0·0001%.

The corrosion resistance of mazak can be improved by a surface passivation treatment after casting. This may take the form of a chromating process carried out by immersing the components in a mixture of chromic and sulphuric acid, which is known by the trade name of 'Cronak'.

Mazak die-castings have the property of shrinking after casting at the rate of about 0·0006 in. per inch. At normal room temperature this shrinking may take several years to complete, but fortunately the phenomenon is greatly accelerated by heat and a normalizing treatment of at least 8 hours at $+70°C$ is recommended after which treatment the casting may be considered fully stable.

It is usually convenient to combine the two treatments of passivating and normalizing so that the overall process becomes: passivate by immersing in Cronak for 10 min then heat to 70°C for 8 hours. The heat cycle serves to dry out the passivation and normalize at the same time.

Spring materials

Various forms of spring are used in the manufacture of instruments ranging from simple coil springs in spring steel, phosphor bronze, etc. where the requirements are not particularly exacting, to the more complex requirements of the spiral springs which are used to provide the restoring force for most instruments. Different instrument manufacturers have used various materials for this purpose, including phosphor bronze, platinum–silver, platinum–iridium, tin–copper and beryllium–copper. These last two have tended to supersede all the other materials with the exception of phosphor bronze, which is still preferred by some manufacturers. The use of most ferrous alloys is prohibited by the fact that for nearly all applications the springs must be non-magnetic.

The tin–copper alloy which contains about 2% tin has been found very satisfactory for precision and laboratory standard instruments. Its specific resistance is low ($2\mu\Omega/\text{in}^3$) and provided that the stress in it is kept to a relatively low value, its mechanical performance is excellent. It suffers from the disadvantage that it is rather soft and easily damaged, so that considerable care and skill are required in handling.

For the less exacting requirements of industrial grade instruments the use of beyllium copper springs has found wide acceptance. This is a dispersion hardenable material which, when correctly heat treated, has a hardness of approximately 700 V.P.N. and consequently will withstand a good deal of maltreatment without significant distortion.

When specifying strip materials for the manufacture of springs, it is usual to specify in addition to the normal composition and mechanical sizes a periodic time requirement which is based on the period of oscillation of a prescribed length of the material when supporting a weight of known moment of inertia. This method is used owing to the difficulty of controlling the material by physical dimensions alone, particularly on the thinner materials. The torque of a spring being a function of the cube of the thickness, it is very difficult to measure the thickness of materials only a few ten thousandths of an inch thick to the required accuracy. An additional difficulty in this respect being that as the strip is usually produced by rolling from round wire the section is not a perfect rectangle, and therefore any calculations based purely on measurements of breadth and thickness are liable to a significant error.

Pivot steels

Many different alloys and materials have been tried for suitability as pivots, ranging from simple carbon steels, through chrome steel, tungsten steel and complex cobalt, chromium molybdenum steels to agate, sapphire and even diamond. These latter difficult and more exotic materials still find application in very special cases but the majority of commercial instruments employ a simple carbon steel. This material is capable of taking a high finish and the hardness after heat treatment is adequate at about 700 V.P.N. Other than its composition (1·0% carbon is typical) its only other property which is important as a raw material is homogeneity and freedom from flaws, cracks, pits, etc. It is normally supplied in the bright drawn condition in short lengths which helps to maintain straightness and freedom from kinks. The main disadvantage with this simple carbon steel is its tendency to rust. More recently an alloy known as Elgiloy or Cobenium has become available which appears to possess all the good qualities of the carbon steel and in addition is stainless. It has the further advantage for use in certain applications that it is non-magnetic. The composition of this alloy is approximately as shown below:

coabalt 40%, chromium 20%, nickel 15%, molybdenum 7%, manganese 2%, beryllium 0·04%, carbon 0·15%, iron, balance.

The cost of cobenium as a raw material for pivots is obviously higher than the equivalent size of carbon steel, but since the amount per instrument is so small and the major cost is in the machining and polishing processes, the increase in cost of a finished pivot is negligible. In a typical example the additional cost of cobenium for a 0·015 in. diameter pivot used on a miniature panel instrument came to only 4s. per hundred or a penny per instrument.

Glass

Glass is used in instruments in at least two important areas. First, and most obviously, it is used as the transparent window through which the instrument is read, and here it is normal to employ a good quality window glass free from striations and inclusions and of a flatness and clarity, which does not cause distortion. For special applications optical quality glass may be used in the form of lenses, optical wedges, etc. Toughened or laminated glass may also be employed where the instrument is liable to be roughly handled or where dramatic temperature changes may occur, such as on liquid oxygen equipment. So called non-reflecting glass which has been lightly etched in hydrofluoric acid is also used sometimes where high incident lighting is liable to cause disturbing reflections.

The other important use for glass is as the material from which jewels are made. The development of glass bearing jewels came about largely due to the shortage of synthetic and natural sapphire during the second world war. It was found that boro-silicate glass (Pyrex) could be moulded to form perfectly satisfactory bearings which were cheaper and easier to produce than sapphire. Glass jewels suffer from one disadvantage compared with sapphire in that they are more liable to fracture under conditions of shock. This led to the adoption of sprung jewels, which when correctly designed, completely eliminate this type of failure. The glass jewels proved to be so satisfactory, that when the old materials again became available they were not generally re-adopted, but were relegated to occasional use where their marginally better performance can justify the additional cost.

Other than correct composition the glass for use in the manufacture of jewel bearings has no particular special requirements. It should obviously be free from inclusions which might cause damage to the moulding tools or surface defects in the finished jewel, but these are not frequent hazards in modern good quality glass.

Magnets

Magnets as applied to electrical measuring instruments are a subject in themselves and complete books have been written about them. The Permanent Magnet Association, to which organization the majority of magnet manufacturers belong, has done a great deal of research on the subject, the results of which are available in various booklets and data sheets published by them and the manufacturers. The largest use of magnets in instruments is in the permanent magnet moving coil type and here the early materials such as cobalt and chrome steels have been almost entirely superseded by alloys containing aluminium, nickel and cobalt. This material is commonly known as Alnico. Different manufacturers use slightly different compositions, but they are generally contained in the following range: aluminium 10–12%, nickel 17–24%, cobalt 10–16%, copper 3–6% with the remainder iron 50–60%. This alloy is extremely hard and brittle and cannot be machined except by grinding. It is normally cast to the required shape and ground only on faces where close tolerances are required.

The same alloy when heat treated in the presence of a strong magnetic field becomes anisotropic, which means that its properties in a preferred direction are improved at the expense of the properties at right angles to the preferred direction. These materials are commonly known as Alcomax in this country, although in America it is called Alnico V.

Both Alnico and Alcomax can also be produced by sintering which allows more complex shapes to be produced consistently without grinding. The performance of the sintered versions is in general slightly worse than that of the equivalent castings, but still quite acceptable, particularly where the process of sintering permits more economic production. The sintering process also lends itself to the production of composite magnets having soft iron polepieces as an integral part of the sintering. The soft iron polepieces can be readily machined to the required contour and holes for fixing provided, which would not be practical in the magnet alloy.

The initial design of a new magnet will normally have been made on the basis of the magnet manufacturers' declared figures for remanence, coercive force and BH max, but the actual flux obtainable in the working gap can only be finally established on the basis of samples. When the required performance has been achieved the problem of measurement and control of production quantities can be tackled. A number of simple methods have been evolved for the testing of production magnets involving fluxmeters, moving iron indicators, where the magnet replaces the normal field coil, and rotary devices but it has been found that the best methods are those in which the magnet under test is located in a magnetic circuit which approximates as closely as possible to the circuit in which it will finally be used. A modified moving coil instrument which has been arranged so that the magnet can be readily changed has been found very suitable for this purpose. Providing that the other constants of this test instrument such as the number of turns on the moving coil and the spring torque are accurately known, then it is only necessary to measure the full scale current with a fully ' raised ' magnet in position. This current can then be directly related to the expected performance of the magnet. By taking a suitable sample quantity from each batch of magnets received, the quality of the batch may be readily assessed. The magnet manufacturers make regular tests on the quality of their materials but the supply of a test instrument to them in which they can check the magnets under actual working conditions is an aid to consistent trouble free performance.

More recently an improved heat treatment process has been developed for cast alcomax which results in an even greater improvement in performance in the preferred direction. Where the shape and size of the magnet is suitable the cooling of the magnet is carefully controlled to encourage the growth of large needle like crystals with their axis in the preferred direction. These crystals exhibit very strong anisotropy. The resulting material is known as a columnar magnet and is even more brittle than ordinary Alcomax. The process is generally only applicable to simple shapes such as rectangular blocks and is only justified where the additional cost of processing is more than offset by the improved performance. Figure 5.2 shows the comparative performance of the various materials mentioned and it will be seen that an additional material known as Platinax has also been included. This is a platinum–cobalt alloy of approximate composition: platinum 77%, cobalt 23%. Due to its high platinum content it is extremely expensive and is normally only used where very tiny magnets with extra high performance are required. Its application to electrical measuring instruments so far has been very limited and the material finds more use in other fields such as electronically controlled watches.

Another group of magnet materials which has found a limited application in instruments in the past few years is the ferrite magnets sometimes colloquially known as ' dirt ' magnets. These materials

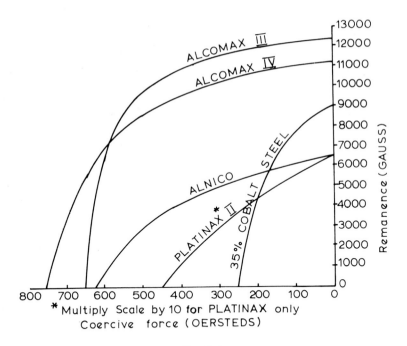

Fig. 5.2.

which are produced by a powder technique have an unusually high coercive force but unfortunately have a rather large temperature coefficient some ten times that of, say, Alcomax which makes their wider application difficult.

The specification and control of magnets is an important function in instrument manufacture, particularly where large quantities of moving coil types are made in a batch or semi-mass production manner. It is essential both to know and control the performance which can be expected from any particular magnet size or type. Failure to utilize the maximum flux available from the magnet is uneconomic since it implies paying for a magnet of larger volume than is really necessary whilst conversely if adequate safety margins are not allowed then an unacceptable number of instruments will fail to reach their design sensitivity.

Plastics

The range of plastics employed in instruments is considerable, ranging from thermosetting materials such as bakelite which has been in use for many years, to the more recent thermoplastics such as the acrylics, polystyrenes, nylons, polycarbonate, etc.

Bakelite, or to give its chemical name phenolformaldehyde was discovered by Dr. Baekeland in the early 1900's and is still widely used. It is well known for its thermal stability and good electrical properties when combined with suitable fillers such as rag, wood, flour, etc. Its chief uses are for instrument cases, field coil supports, small insulators, etc. A variety of similar resins such as urea formaldehyde have since been developed extending the use of the thermosetting resins even further. The formaldehyde resins are moulded by a combination of heat and pressure and after the appropriate curing period in the mould, the material polymerizes, yielding a part which faithfully reproduces the shape and finish of the mould. Quite a high finish may be produced from a polished mould since a thin skin of the pure resin is normally formed on the surface. This skin, if left unbroken, is largely impermeable to moisture and resistant to attack by many of the common solvents, weak acids and alkalis. The electrical properties of the material are good under normal atmospheric conditions, but if the skin is broken and the part is left in contact with free water, the water permeates the granular type bulk material, degrading the performance to an extent depending on the type of filler employed. It is generally a hard, somewhat brittle material, unsuitable for highly stressed parts, but retaining great popularity on account of its cheapness and good all-round performance.

Variants on the original phenol formaldehyde have been developed for improvement in particular properties, but where more extreme conditions of temperature, solvent resistance, electrical or mechanical properties are to be met, then the thermosetting plastics tend to be replaced or superseded by one or more of the enormous and ever increasing range of thermoplastics.

The thermoplastics differ from the thermosetting plastics, first in that, as their name implies, they become liquid with the application of heat. The temperature at which this change of state occurs varies very widely from one material to another, and this, coupled with the wide range of their other properties, enables only the broadest generalizations to be made here concerning the characteristics of the more common groups.

Cellulose plastics

Cellulose nitrate and cellulose acetate were two of the earliest available thermoplastics. Both required slightly special techniques for moulding, in the case of the nitrate, due to its instability on heating and the acetate because it is affected by water, Cellulose acetate butyrate which is a more recent version is less affected by water and has consequently found wider application. It is a tough transparent material having a heat distortion temperature of 60°C to 70°C.

Polyamides

These materials were first introduced as fibres (the well-known nylon) but have since been developed as moulding materials. The basic material is a hard horny substance with a heat distortion temperature of 70°C to 80°C but a wide variety of different modified forms are available with low water absorption and varying mechanical properties. Nylon moulds well but requires particularly well-made tools, since at moulding temperatures its viscosity is very low and it will readily penetrate into quite small gaps. It has been successfully used for small gears and bearings and can be run with little or no lubrication providing dirt is carefully excluded. It also has good electrical properties and is widely used with or without the addition of glass fibres and other fillers.

Methylmethacrylate

This material is more widely known by one of its trade names such as ' Perspex ' when used in sheet form or ' Diakon ' as a moulding material. It is widely used for its great clarity and has found considerable application in instruments in recent years as the material for the variously shaped clear fronts or covers employed by most manufacturers of miniature instruments. Its clarity when carefully moulded compares favourably with that of glass, although it cannot compare with glass for scratch and solvent resistance. Its softening temperature lies between about 60°C and 90°C depending on the grade. Its electrical properties are excellent, although this tends to be a disadvantage in its use for instrument covers, due to its tendency to develop high static charges which are difficult to disperse. Many treatments have been developed, mostly aimed at lowering surface resistivity and hence allowing static charges to leak away. The majority of these treatments can unfortunately only be classed as temporary or at best semi-permanent so that the use of the material in close proximity to light mechanisms such as the moving systems of electrical measuring instruments, should always take due account of this possible difficulty.

Perspex or Diakon is also available in a wide range of translucent and opaque colours and finds considerable application in optical components such as lenses and diffusers as well as in more decorative components. It also is much used whenever very high insulation resistance is required in measuring apparatus of many kinds. It is a rather brittle material and whilst suitable for use under moderate compressive stresses should never be used for components subjected to significant tensile or shear stresses. Careful control of mould design and moulding conditions are necessary to avoid stress cracking, particularly if the component is likely to come into contact with solvents, either during manufacture or subsequent use.

Polystyrene

Although of different chemical composition, this material in its unmodified form is similar in many ways to methylmethacrylate. It, too, has good clarity and excellent electrical properties. It is somewhat more brittle than 'Perspex' and hence more liable to stress cracking. It is generally cheaper than Perspex and is often used where cost is particularly important. It is also available in various toughened, light stabilized and other grades and forms the basis of a number of modified forms such as A.B.S. (acronytile-butadrene-styrene) and others whose properties are too varied to be described here.

Polycarbonate

This is a material of American and German origin which has found wide application in the electrical industry, particularly on the continent, due to its excellent non-tracking properties and good mechanical strength. It is more commonly known by its trade name of 'Makralon' and its introduction to uses in electrical measuring instruments is of fairly recent origin. In its natural form it is transparent with a rather yellowish tinge but a range of colours is also available. It has excellent dimensional stability, great toughness and a wide temperature range, the softening temperature being about 135°C. Its solvent resistance is only moderate but adequate for most instrument applications. Careful control of the moulding material is required due to its tendency to absorb moisture whilst in the powder or granular form. Components moulded from powder containing moisture are brittle and easily broken, but this condition is not readily detectable except by destructive tests.

Material selection

The selection of the most suitable material from which a component is to be manufactured is influenced by a number of factors. The relative importance of which will vary depending upon the component and its application. Some of the more important factors are mentioned below, although no claim is made that the list given is complete. Every engineer faced with this problem is well advised to give due consideration to whether there are any other factors which should be taken into account in his own particular case.

(a) Function

This is normally the first consideration and in many instances will enable the possible range of materials to be greatly narrowed. For example, must it be a conductor or an insulator. If the former is the case then the choice is generally restricted to metals whilst if the latter, then plastics, glass, ceramics, mica, etc. would be the choice.

These simple considerations enable the range of possible materials to be narrowed down to a particular group or groups and the addition of further functional requirements such as, temperature range, magnetic properties, hardness, tensile strength, solderability, optical properties, etc. will enable the choice to be narrowed still further. Several commercial systems such as that supplied by Materials Data Ltd. are available to aid the engineer in what may be for specialized components a rather tedious task. In the materials Data system, the various properties of all the likely metals and their alloys are listed on cards, one card to each metal or alloy. Each card has an index number. On special master cards all the properties are grouped and by a system of overlay transparencies are stacked over the appropriate master card. The reference numbers of only those materials fulfilling all the required parameters can then be read off. A similar system for plastics is also available.

In some instances the above procedures will result in the selection of one material only, but more frequently several materials will appear to satisfy the requirements equally and even further factors must be taken into account before a final selection is achieved. The factor which is usually next in importance to function is:

(b) Cost

Besides the simple consideration of costs per cubic inch or cost per pound etc., this also must involve less tangible factors such as the cost of introducing, say, a new moulding material, the possible quantities required, the availability of suitable machinery for manufacture, the tooling cost, say, for a moulding etc.

Finally, it must be remembered that the choice of a material for any one component cannot be made in isolation; its compatability with the materials of adjacent components, must be taken into account. The indiscriminate mixing of different plastics in contact or adjacent to one another may well cause trouble. For instance, the plasticizer in certain grades of p.v.c. will cause discoloration and even cracking of, say, diakon or polystyrene. In the author's experience there is not a great deal of published information on this subject and in cases of doubt the only solution is carefully controlled experiments to determine the degree of interaction if any. The most likely source of trouble are those materials such as p.v.c., rubber, etc. which contain significant amounts of plasticizer or other free chemicals such as sulphur.

Where the adjacent materials are metal, different consideratons apply. In the presence of quite small amounts of atmospheric moisture the junction between dissimilar metals may well become a tiny battery giving rise to an electrolytic action, and consequent corrosion. For normal industrial use this effect is negligible if the materials in contact are less than about 0·5 volt apart on Table 5.1, although for use under more humid conditions such as are found in the tropics, then it is recommended that the permissible separation be reduced to 0·2 volt or less.

Material	Volts
Magnesium and its alloys	−1·60
Zinc and its alloys:—	
Zinc die-casting alloy, B.S.1004	−1·10
Zinc plating on steel, D.T.D.903	−1·10
Zinc plating on steel, chromate passivated, D.T.D.923	−1·05
Galvanized iron, B.S.729	−1·05
Tin-zinc alloy (80/20) plating on steel	−1·05
Cadmium and its alloys:—	
Cadmium-zinc solder, D.T.D. 221	−1·05
Cadmium plating on steel, D.T.D. 904	−0·80
Aluminium and its alloys:—	
Aluminium alloy casting, B.S. L5	−0·90
Aluminium alloy casting, D.T.D. 424	−0·80
Aluminium alloy castings, B.S. L33, L51	−0·75
Wrought aluminium alloy coated aluminium alloy, D.T.D. 687	−0·90
Wrought aluminium, B.S. L4, L16, L17, L34, T9	−0·75
Wrought aluminium coated aluminium alloys, B.S. L38, D.T.D. 546, 610	−0·75
Wrought aluminium alloys other than duralumin	−0·75
Duralumin type alloys, B.S. 395, 396, L1, L3, L39, T4, D.T.D. 364, 464, 603, 646	−0·60
Irons and steels:—	
Non-stainless:—B.S. S2, S6	−0·75
B.S. S3, T1, T45, Grey cast iron	−0·70
Stainless:—12% chromium, B.S. S61, S62, S85, D.T.D. 161, 203	−0·45
High chromium, B.S. S80, D.T.D. 60, 146, 185	−0·35
Austenitio, B.S. S110, T55, T58, D.T.D. 166, 171	−0·20
Lead and its alloys:—	
Lead	−0·55
Lead-silver solder (2½% Ag)	−0·50
Tin and its alloys:—	−0·50
Tin and its alloys:—	
Tin-lead solders, B.S. 219, Grades A and B	−0·50
Tinned steel, B.S. S20	−0·50
Tin plating on steel, D.T.D. 924	−0·45
Chromium:—	
Chromium plating, 0·0005 inch on steel	−0·50
Chromium plating, 0·00003 inch on nickel-plated steel	−0·45
Chromium (massive, 99%)	−0·45
Copper and its alloys:—	
Brass, B.S. 265	−0·30
Brass, B.S. 249; Gunmetal, B.S. 383	−0·25
Copper, B.S. 899; Copper-Beryllium, T.C.M. Cu-Be 250; Brasses, B.S. 250, D.T.D. 283; Bronzes, B.S. 407, D.T.D. 412; 'Nickel-silver', B.S. 790; Cupro-nickel (70/30)	−0·20
Nickel and its alloys:—	
45% Nickel alloy, D.T.D. 237	−0·25
Monel metal, D.T.D. 10; Monel X	−0·15
Nickel plating on steel, B.S. 1224 (Ni 8S Category)	−0·15
Silver and its alloys:—	
Silver solder, B.S. 206 Grade C	−0·20
Silver; Silver plating on copper, Silver-gold (90/10) alloys	±0
Rhodium plating on silver-plated copper	+0·05
Platinum	+0·15
Gold (Assay grade)	+0·15

Table 5·1. Potentials against a saturated calomel electrode in sea-water at room temperature.
(values to nearest 50 millivolts)

This problem of electrolytic action may often be solved by the use of suitable electroplating on one or both components which leads us on to the general consideration of:

Finishes

The range of finishes available is so vast that the detailed treatment of it is a subject in itself, and the remarks given here will be restricted to those finishes most commonly encountered in the manufacture of indicating instruments. Finishes are generally applied for one or more of the following three reasons: protection, appearance or function and in this connection a finish is defined as a surface coating or coatings applied to or caused to appear on the surface of a component after it has been fabricated. It does not include the application of lubricants or temporary surface films applied purely for the purpose of protection during subsequent manufacturing operations.

For metal components the most commonly used finish is some form of electroplating. For the ferrous materials the primary need is usually protection against rusting and for this purpose such platings as nickel, tin, tin–zinc, tin–nickel or cadmium are commonly applied, thicknesses of up to about 0·0005 in. being necessary to give adequate protection against rusting in normal industrial atmospheres. Where high surface finish or reflective properties are required polished nickel or polished chromium are usually employed and in such cases it is normally necessary to polish the surface of the component before the plating is applied.

Very pleasing appearance may also be obtained by the application of one of the very wide range of paint finishes now available. By a suitable choice of paint with or without previous plating a good degree of rust protection can also be achieved. A fairly recent introduction is painting by an electrophoretic process in which the paint is deposited on a conducting surface from an acqueous suspension in a manner similar to electroplating. This is a very convenient method of painting awkwardly shaped components and results in a very consistent coating thickness without drips and runs.

For non-ferrous metal components the use of plating is equally applicable although the corrosion prevention is often not so necessary and the plating thickness can be reduced to 0·0001 in. or less, purely for appearance. Painting again is often employed particularly for items such as instruments, scales, pointers, etc. which are usually non-ferrous.

Aluminium and aluminium alloy parts can only be plated with some difficulty and are usually anodized or painted. Anodizing of aluminium is again a process similar to electroplating which results in the formation on the surface of a thin layer of hard aluminium oxide which, depending on the particular process, can be used as a very effective insulator. Fresh anodizing also has the property of being readily dyed, and very attractive finishes can be achieved in this manner.

In instrument manufacture good solderability without the use of highly active fluxes is often required and for most ferrous and non-ferrous materials except aluminium a very thin gold plating has been found very effective. It has the particular advantage that good solderability is maintained for long storage periods whereas most other plating finishes whilst readily solderable when fresh tend to lose this facility fairly quickly when stored. A very convenient way of applying the thin gold plating required for this purpose is by immersion or electroless plating. Special solutions are available for this purpose and do not involve any electrical connection, being purely chemical in action. Plating thicknesses of as little as 0·00001 in. have been found effective so that the cost is only a fraction of that normally associated with gold plating.

The plating of certain plastic materials may also be carried out by the immersion process, usually for decorative effect, but sometimes to form a conducting film for a subsequent electroplating process.

Conducting or reflecting metallic films may also be deposited on plastics by vacuum deposition. Typical examples of the use of this process being the deposition of aluminium on the inside of a bakelite instrument case to form an effective electrostatic shield or the deposition of aluminium on to diakon to form a mirrored surface in an optical system.

In conclusion it cannot be over-emphasized that for the instrument design engineer the subjects of material selection and finishes are worthy of considerable attention, because herein lie many of the answers to today's pressures for reduced cost and enhanced performance. The rapid advances in materials, particularly plastics, enable designs to be considered which would have been quite impractical or prohibitively expensive only a few years ago. This is not to assume that because it is new it must be good, but to suggest that as new materials and processes become available, they should be carefully assessed to see if they open up new design possibilities.

CHAPTER 6
presentation

RELATED to electrical indicating instruments the term presentation is rather loosely used by manufacturers and users alike to refer to the general shape and appearance of an instrument, particularly those parts which are visible when it is installed and in use. Especially for panel mounting instruments the general shape tends to be a matter of fashion; being largely dictated by the current trend in equipment design. Thus design has followed demand through such phases as 'the round look', 'the square look', 'the rectangular look', 'the curved look', 'the sharp corner look' and there is little reason to suppose that these various phases will not continue as long as instruments are required in sufficient quantity to justify the necessary design and manufacturing effort to keep in tune with the latest equipment fashion.

However, regardless of shape, size or type all indicating instruments have two essential features in common, a scale and an index or pointer showing the magnitude of the quantity being measured. After all questions of fashion have been forgotten it is the scale and pointer which the user is going to look at most frequently. These parts, therefore, merit a good deal of design attention to ensure that they enable the required reading to be made as quickly as possible with the least possible likelihood of error. A good deal of research has been carried out by various bodies on the ergonomics of making instrument readings. Although the resulting recommendations are by no means unanimous it is the general consensus of opinion that the first consideration is whether the instrument is required for accurate reading from a short distance or for less accurate reading at a greater distance. The less common requirement to combine both features will be dealt with separately. The necessity to classify instruments according to the reading distance results from the physiological fact that the human eye can only discriminate between two points when the distance between them subtends a certain minimum angle at the eye. Statistical measurements show that for the average human eye this angle is about $2'$ of arc. It follows from this that it is pointless to have divisions smaller than the eye can discriminate from the normal reading distance. The same research also showed fairly conclusively that the eye can split a division into two, five or even ten parts just as accurately without the aid of very small subdivisions. In addition there is less likelihood of error if the eye is not confused with the problem of counting a large number of divisions from the nearest numbered mark. For much the same reasons the recent trend is to remove all unnecessary markings from the instrument scale so that the eye is not distracted from the primary object of making a quick and accurate reading.

The British Standard Recommendation 3693 sets out in some detail all the suggested scalings based on the principles briefly outlined above and this document gives a sound basis for normal scale design. It does not purport to cover all the special cases which may arise but even in such instances its broad recommendations form a sound starting point.

Some examples of typical scales are given below, in fig. 6.1 together with a table showing the relationship between division size, accuracy, etc., related to viewing distance. The dual purpose scale shown is a compromise intended for close or distant viewing, the fine divisions being used for close reading to best possible accuracy and the thickened divisions when less accurate readings are taken from a distance.

Scale materials and processes

The scale material for most modern instruments is metal, brass, aluminium or even steel having been variously chosen by different

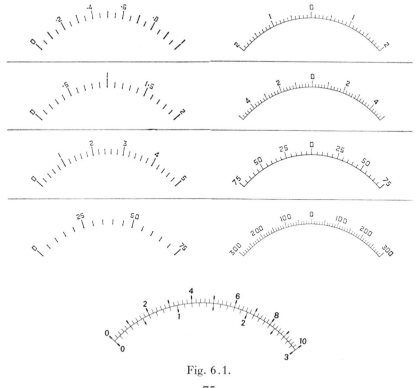

Fig. 6.1.

Rdg distance (ft)	Scale base length (inches)
2	1·7
3	2·5
4	3·3
6	5·0
8	6·7
10	8·3
12	10·0
15	12·5
20	16·7
25	20·8
30	25

Formula relating above is $L = \dfrac{D}{14 \cdot 4}$ which is based on experimental evidence and relates to an observation accuracy of $\mp 1\%$ F.S.D.

manufacturers. Plastics have also been used in certain instances, but generally the problems such as static, associated with these materials have limited their application to special cases, such as for back illumination, where their advantages outweigh the disadvantages.

For maximum readability a high contrast is required between the scale plate and its markings, consequently the normal arrangement for most industrial instruments is a white dial with black markings. The white background is produced by spraying the scale plate with a white enamel and the markings are put on subsequently by one of several methods. For high quantity industrial grade instruments the markings are applied by silk-screen printing or by an offset process, the required scale shape having been determined by initial measurement on a representative batch of instruments. A refinement on this arrangement is the production of two or more scales differing only slightly in scale shape. Each individual instrument is then checked to determine its scale shape and the appropriate scale fitted. If this refinement is still insufficient to obtain the required accuracy, or if the quantity involved is small, the old fashioned method of individually hand drawn scales is still employed. Such scales are expensive due to the time and skill required to produce an acceptable looking scale and their use tends to be limited to the precision grade and laboratory standard instruments where the additional cost can be justified.

Various ingenious machines have been devised to enable the accuracy of a hand drawn scale to be achieved whilst eliminiating the manual skill required. In general they have only been worth while where the quantity of one particular instrument type involved has been large. Consequently, they have not been widely adopted, as most instrument manufacturers have a demand for a large variety of scales but relatively few of any one type.

Various photographic methods for producing scale markings have also been developed. Some of them are capable of producing a very high quality scale, the clarity and sharpness associated with photographic processes being fully realized.

The choice of method to be employed in any particular instance is a complex one involving both technical and costing considerations, varying from one company to another and quite impossible to solve except by the expert on the spot. Pointers also have to be designed to be compatible, with one or other of the two main scale classifications. For close accurate reading the pointer tip is often made into a knife edge having a thickness about the same or a little less than the thickness of a division line on the scale. For extreme accuracy the addition of an anti-parallax mirror facilitates easy reading, the correct reading position for the eye being when the reflection of the knife is directly under the knife and therefore not visible. To aid further the attainment of this condition the upper and lower edges of the knife edge are often painted in different colours such as red for the top and black for the underneath. Any lack of coincidence of the pointer and its image is thus more easily distinguished.

With the fairly low powered and delicate mechanisms used in most indicating instruments it is obvious that the pointer must be light in weight. In spite of this it must also be rigid enough to withstand reasonable handling during assembly and the normal overloads and jars it will be expected to meet in service. Various different designs have been tried by different manufacturers but the majority are based either on the use of thin aluminium alloy blanks or thin walled tubes. Pointers made from tubing can be made very light as well as rigid without undue difficulty, but careful ribbing or curving of the blanked pointer is necessary if the weight is to be kept to a minimum without rendering the pointer too fragile for its purpose.

For larger instruments such as 6 in. precision grade or 12 in. laboratory standards the simple tube construction is often replaced by a more rigid construction known as the truss. A sketch of this construction as used by Sangamo Weston is shown in fig. 6.2.

This construction being effectively formed from two triangles is very rigid without adding excessively to the weight or moment of inertia.

Whenever any of these pointers are specifically required for distant reading it is often the practice to fix a disc of thin aluminium near its end.

To improve the visibility of the pointer against the normally white scale plate the pointer is usually blackened. This can be done by black anodizing if the pointer material is aluminium, but needs considerable care to give a good matt black. More commonly the pointer is painted with a matt or semi-matt black paint. Considerable care is necessary in choosing this paint, particularly for the more sensitive instruments having low torques as any tendency for the weight of the paint to vary, for instance due to hygroscopy, will result in variable balance errors.

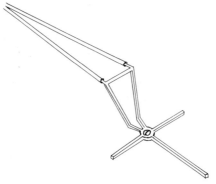

Fig. 6.2.

Variations of a few hundredths of a milligramme could well lead to balance errors greater than 1% of scale length.

Anti-parallax devices

Mention has already been made of anti-parallax mirrors. These are used almost exclusively in conjunction with some form of knife-edge pointer, the correct use of the mirror relying on the fact that a small deviation of the eye from the correct reading position causes the reflection of the underside of the knife edge to become visible. The mirror surfaces are produced in different ways by different manufacturers, some consisting of glass mirrors, some highly polished metal mirrors. A recently introduced material is a thin plastic adhesive tape having an aluminized surface. When pressed firmly on to a flat surface this material forms an effective mirror.

For various reasons such as non-acceptable reflections or the necessity to use pointer types not incorporating a knife edge the so-called mirror scale is not always employed, although an anti-parallax arrangement is still required. In these instances the construction known as a platform scale is often used. This is illustrated diagrammatically in fig. 6.3. The essential feature is that the portion of the scale plate carrying the scale markings is raised and the pointer tip is arranged to swing in the same plane as the raised portion and as close as possible to the end of the scale marks. In this way even at quite large

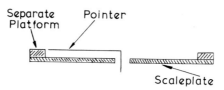

Fig. 6.3.

angles to the normal there is no apparent relative movement between the pointer tip and the scale mark. In the more expensive instruments the platform scale is cut from the solid or made in two parts so that a cross section of the scale is as shown in fig. 6.3. If the radial clearance between the pointer tip and the platform are kept down to say 0·010 in. to 0·020 in. this construction is very effective. A slightly less effective but often acceptable and less costly construction is that shown in fig. 6.4, where the scale is made in one piece and the platform produced by a forming operation. Due to the necessary corner radii the pointer tip cannot be made to run quite so close to the ends of the scale marks and the possibility of a small reading error is increased, although the method is a considerable improvement on having no anti-parallax feature at all.

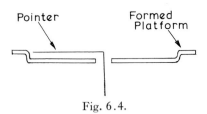

Fig. 6.4.

No article on instrument scales would be complete without mention of the scaling normally employed on laboratory standard instruments. It is often known in the trade as the ' vernier scale ' although it is in fact not a vernier at all. To the author's knowledge no name has ever been given to it which correctly describes its function. As will be seen from fig. 6.5 it consists of a number (usually six) of concentric and equally spaced arc lines spanning most of the length of a division line. A very fine diagonal line is drawn from the base of one division line to the top of the next where they intersect the inner and outer arc lines. By this means the distance between adjacent division lines is accurately divided into, say five equal parts, the boundary of each part being the intercept between the concentric arc line and the diagonal.

The rather elaborate and costly method can normally only be justified on the most accurate instruments such as laboratory standards necessitating as it does great care in the manufacture of a rather long knife edge

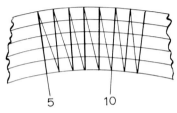

Fig. 6.5.

on the pointer which must be very straight and square since readings are taken at different points along its length. The care necessary may be better appreciated when it is realized that on a laboratory standard of say, 12 in. scale length offered at an accuracy of $\mp 0.1\%$ of full scale deflection all errors must be included within a pointer movement at the tip of only twelve thousandths of an inch.

CHAPTER 7
manufacturing techniques

THE aim of all manufacturers, including instrument companies, is (or should be) to produce the right goods at the right time with the right quality at the right price.

For the purposes of this chapter it has been assumed that all the necessary prerequisites such as good marketing, good tooling and good manufacturing facilities have been achieved. It is further assumed that the case of a company is considered which manufactures at least a majority of its own piece parts, so that the different problems associated with the purchase of large quantities of piece parts from sub-contractors will not be considered in any great detail.

Some of the problems of material selection and control have been dealt with in a previous chapter so that here we shall concentrate primarily on the manufacture of piece parts and the various stages they will pass through as sub-assemblies and assemblies before the instrument of which they form a part is ready for shipment to a customer.

One of the prime requirements for all instrument parts is that by the time they reach the assembly stage they shall be very clean. All instrument types are very susceptible to quite small pieces of foreign material in the wrong place. With high magnetic fluxes which occur in small working gaps, magnetic particles are particularly troublesome since they tend to get drawn into the instrument's most vulnerable areas.

The overall cleanliness of the finished instrument is not, as may be thought, just a matter of good cleaning operations on parts after manufacture but also involves good design to avoid constructions and operations which create and harbour dirt in its widest sense and very careful control of manufacturing operations and tooling.

From the design point of view blind holes, particularly threaded ones, and any construction which leaves rough or inaccessible crevices are to be avoided.

During manufacture it is essential that a number of precautions are observed. First, wherever possible clean and dirty operations should be segregated into different areas so that clean parts are not contaminated by swarf etc., produced during dirty operations, such as drilling, grinding, milling, etc. Also, it should be ensured that all tools such as drills, taps, cutters and so on are kept properly sharpened so that ragged burrs which may become detached at a later stage are not left on the piece parts.

After all machining operations have been finished most piece parts will need to be cleaned to remove oil or grease and any loose particles resulting from the machining operations. On fairly smooth parts a degreasing operation may be all that is necessary and here again care is required not to contaminate parts by picking up dirt in this operation. If the parts are not very greasy the cleanest method is by vapour degreasing in which the parts are suspended in hot vapour from a solvent such as trichlorethylene. The vapour condenses on the relatively cold pieces and runs off carrying the oil or grease with it. This process is normally allowed to continue until the parts are hot enough for the condensation to cease.

For parts having a more complicated shape other additional cleaning operations may be necessary to remove any trapped or lightly adhering particles. Such processes as centrifuging, tumbling or ultrasonic cleaning are used as appropriate in such instances although in the most stubborn cases it may be found essential to employ tedious hand cleaning processes. Ferrous parts are often rather difficult to clean owing to the tendency of some machining operations to magnetize them slightly, with the consequence that the machining swarf is likely to cling. In such circumstances a thorough demagnetizing treatment has often a proved effective when applied before one or more of the normal cleaning methods. A common failing after careful cleaning of piece parts is to return them to the dirty container in which they arrived. It must be noted that the cleanliness of the containers is just as important as the cleanliness of the parts. Consideration is also required of the materials from which containers are made. Wood, paper, cardboard, etc., are fibrous and if possible should be avoided, as during handling and general wear and tear, small particles and splinters may break away and contaminate the cleaned parts. Plastic containers are less liable to this failing but unless treated with a suitable antistatic solution are likely to pick up a static charge and attract airborne dust etc. Metal or glass containers are suitable for most parts and all containers should have lids if likely to be stored for any length of time or be transported through dirty conditions. One of the most satisfactory arrangements which overcomes many of these problems of cleanliness is the system in which all piece parts and sub-assemblies must pass through a cleaning process in order to get into the final assembly or clean area.

For the final assembly of all types of indicating instruments good clean working conditions are essential. This implies clean working surfaces such as Formica on benches, smooth floors, walls and ceilings, adequately filtered air conditioning and general cleaning routines. Such facilities are an obvious requirement but what may not be so readily appreciated is the ease with which the effectiveness of a clean area can be ruined by a little thoughtlessness. Apart from magnetic particles which are largely removed by adequate piece part cleaning the biggest cause of 'dirty' instruments is lint or hairs. The sources from which these may arise are legion, ranging from textile fibres from clothing or overalls to human hairs, brush hairs and particles from paper etc. If possible, all paper such as orders, specifications, schedules, etc. should be eliminated from the clean area altogether. If this is not completely possible then only lint free paper should be used or each piece of essential paper contained in a transparent plastic envelope. The maintenance of an effective clean area requires constant vigilance on the part of every person using the area, but the effort involved has been shown time and time again to be well justified by the improvement in general quality which can be achieved.

Certain assembly operations such as soldering are difficult to eliminate from the clean area altogether and being classed as 'dirty' operations need special attention. Various techniques such as vacuum lines to remove particles of scale and solder have been adopted, together with a routine of removing soldering irons for cleaning regularly once or twice per day. The provision of special soldering iron rests into which the iron is placed when not in use have also proved effective.

Another common source of foreign particles is the small tools such as screwdrivers, box-spanners, etc. which are used during many operations of final assembly. These should be inspected at regular intervals to ensure that they are in good condition and also that the correct sized tool is used at all times. Too small a screwdriver for instance, can create considerable trouble by bruising screwheads such that small particles are removed. Such particles may fall unseen into the working gaps causing havoc, as what is colloquially known as friction at a later stage when the instrument is on test.

It will be seen from the foregoing remarks (which by no means cover all the possible trouble spots) that the required cleanliness can only be achieved by a great deal of thought, effort and attention to detail and that only with goodwill and cooperation can the necessary standard once achieved be maintained. It has been said in this connection that a clean area is not so much a place as an attitude of mind. Whilst this may be somewhat of an exaggeration it contains a good deal of truth worthy of consideration by any individual or company faced with the problem of setting up such an area. Without the cooperation of its occupants the most expensive clean room will rapidly be rendered completely ineffective. Whereas with their cooperation even a modest installation will amply justify its cost.

The primary requirement for a clean area is during all stages of what may be termed final assembly. As previously mentioned all piece parts arriving at this stage should have been thoroughly cleaned and packed into clean containers so that no dirt is carried into the clean area.

The following description applies primarily to the assembly of small moving coil instruments conforming to British Standard Industrial Grade but many of the techniques involved apply equally to larger precision grade instruments, although these tend to be made in smaller quantities and therefore receive more individual attention than the high quantity miniatures.

Once the necessary parts are available the first operation is the assembly of the moving element consisting of the moving coil, the pointer and the springs. Different manufacturers use slightly different constructions but a typical Sangamo Weston design is shown in fig. 7.1.

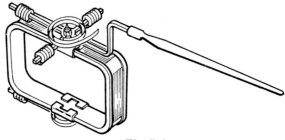

Fig. 7.1.

The moving coil arrives complete with pivot bases. The pointer is slipped over the top pivot base and located on the two flats on the pivot base. Next the springs are put on, the inner end of the spring being carried on a small terminal which also slips over the pivot base. The outer ends of the spring are positioned to give the required free zero position (left hand, centre, right hand or intermediate) and finally the pivot base nuts are screwed down the threaded portion of the pivot base to clamp the springs and pointer firmly to the pivot bases. The two ends of the moving coil have already been soldered, one to each pivot base, so that the mechanical clamping due to the pivot base nut also provides an electrical connection between the coil ends and the springs. All these operations are carried out on a suitable assembly fixture such as that shown in fig. 7.2. This fixture is also used for the last part of the assembly of the moving element which is the insertion of the pivots. A hardened steel pivot is carefully inserted into the pivot punch shown alongside the fixture. The protruding shank of the pivot is then inserted into the hole in the pivot base and the punch is pushed steadily but firmly down until it is stopped by the top face of the pivot base. The punch is then gently withdrawn. The pivot is held in position in the pivot base by being an interference fit in the hole. This method of pivot insertion ensures that the pivot projection is controlled to a very close tolerance. This is an essential feature of most Sangamo Weston designs and when taken in conjunction with other closely controlled dimensions ensures that the pivot cannot come out of its jewelled bearing except by applying sufficient force to cause considerable mechanical distortion of one or more of the component parts.

After the pivots have been inserted the moving element is inspected for general cleanliness, spring position and flatness, straightness and

Fig. 7.2.

squareness of pointer, etc., before passing on to the next operation, which is ' swinging ' the movement. By this is meant the operation of assembling the moving element together with the necessary parts of the magnetic circuit so that the pivots are suspended in the jewel screws and the movement is free to swing. Whilst this operation is basically the same for all movements the actual details of the operation vary widely according to the construction of the particular movement design. In the latest Sangamo Weston internal magnet movement this operation is largely automated. The appropriate parts are merely laid in position in a special assembly machine and by a push button activation a sequence of pneumatically driven pistons assembles and secures the assembly, delivering a completed swinging mechanism to the operator. This completed mechanism then only requires a simple soldering operation to connect the outer ends of the springs to their appropriate abutments. The movement is then inspected, rough balanced and placed in an individual movement box which has a lid to keep out any dust and also carries the identification of the movement and its order in code form. The movement then travels on to the next operation which is ' raising ' or magnetizing and scale determination.

Magnet raising is carried out by the modern capacitor discharge technique which has almost entirely displaced the old methods requiring high d.c. currents in cumbersome coils supplied by large d.c. generators. The modern unit is fed from a normal a.c. power socket and the internal transformer rectifier unit supplies a d.c. voltage of up to about 1000 volts which is used to charge a large capacitor bank. When the capacitor bank is fully charged, which takes only a few seconds, it is discharged into an an iron-cored coil on which polepieces appropriate to the movement concerned are mounted. The capacitor discharge gives a very high peak current enabling magnetizing powers of 30 000 ampere turns or more to be readily achieved. Figure 7.3 shows a typical magnetizing jig suitable for the internal magnet design referred to above.

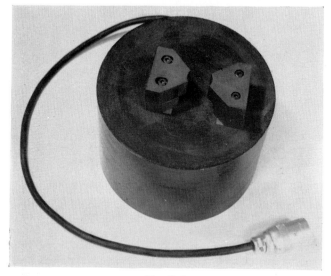

Fig. 7.3.

After raising, the movement is checked for electrical sensitivity which must fall within certain limits according to the requirements. This ensures that the correct springs and moving coil have been fitted. The magnet is then aged leaving the full scale sensitivity 4% to 5% high of the required final value. The movement is then checked for scale shape to determine to which of the two or three preprinted scales it most nearly corresponds. The identification letter of this scale is recorded and the movement replaced in its box. The box containing the movement is put into an oven and baked for a period of 24 hours at 70°C. This process has been found to normalize any parts which have been subjected to slight strain during assembly and also finally to remove any traces of moisture or solvents which remain in the pointer or moving coil.

When the movement has cooled to room temperature after its baking period it is returned to the assembly line for completion. During the baking period the appropriate scale has been prepared; this is now fitted, the jewels finally adjusted and the movement finally balanced as closely as possible. It is then given a complete electrical check and its case finally secured, before being passed to a final mechanical and electrical inspection. On satisfactory completion of all the inspection tests the instrument is sealed with a company seal and is ready for shipment to the customer.

Several references have been made in the preceding part of this chapter to inspection operations. In the manufacture of any articles in large quantities an inspection department has a very important role to play and the production of instruments, particularly miniatures, is no exception. When small quantities of special instruments are individually made the quality of the finished item is very much dependent on the skill and experience of the instrument maker concerned, who can be relied upon to take sufficient pride in his own workmanship to ensure a high quality product. When instruments are produced in larger quantities for sale in a competitive industrial market it is largely the inspection department which is responsible for maintaining the quality of the product, once a satisfactory design has been put into production. They are usually responsible for checking the product at all stages from the receipt of raw material right up to the shipment of the finished item.

During the manufacture of piece parts the application of an efficient system of quality control can save much wasted effort occurring at later stages of assembly by ensuring that the percentage of good usable piece parts is approaching 100%, whilst efficient inspection during assembly will also ensure a minimum of wasted effort. To enable all the various inspections to be carried out efficiently considerable attention is given in the modern instrument industry to the provision of suitable gauges, fixtures and optical aids to assist the inspector to perform his function.

Electrical inspection presents similar problems and considerable use is made of electronic devices for checking and supplying the wide range of both a.c. and d.c. sources required to energize instruments and their auxiliary devices. A comprehensive system of regular checking and cross checking is required to ensure the maintenance of accurate electrical standards. These are usually derived from equipment such as standard cells, d.c. potentiometers standard resistances, etc., which are checked and issued with a certificate by the National Physical Laboratory. By the careful use of such standards in comparison with the working standards the accuracy of the latter is regularly cross-checked.

CHAPTER 8
extension of range

MOST of the instruments which have been discussed are restricted in self-contained ranges to upper limits of a few tens of amperes and a few hundreds of volts. For a number of measurements it is necessary to extend this range by a factor of up to a hundred times or more. To achieve this extension various devices such as series multipliers, shunts, voltage transformers and current transformers are used. Since the design and use of such items may have considerable influence on the accuracy of the measurement some consideration of the factors involved will not be misplaced.

(a) Series multipliers

These may be made suitable for both a.c. and d.c. measurements and a number of different constructions have been utilized. For greatest resistance accuracy they are normally wire wound in a material having a low temperature coefficient of resistance such as constantan, minalpha or nickel–chromium. For resistance stability it is essential that the winding should be as strain free as possible and both bobbins and flat cards are used as formers. The latter have the added advantage that no special winding is necessary for use on a.c. and their heat-dissipating properties are excellent. The use of card resistors of this type is usually restricted to the high accuracy, high current applications such as multipliers for the potential ranges of air-cored dynamometers, where the relatively large space required can be justified.

For lower current multipliers such as are required for moving coil voltmeters considerable use is made of metal-oxide or metal film resistors whose characteristics are perfectly satisfactory for all but the very highest precision, when wire wound resistors are still generally employed.

For high voltage multipliers, say above about 2000 volts, and for currents of up to a few milliamperes metal–oxide or metal film resistors are generally used and to improve the insulation they are often ' potted ' using one of the well-known materials such as epoxy resins. Using this type of construction ranges of at least 20 kV are readily achieved.

Whilst the manufacturer will provide the correct type and construction of multiplier in accordance with the range and accuracy required, it is worth while to note some of the design pitfalls which must be avoided: Wire wound resistors must be constructed in such a manner that the wire is as strain-free as possible, otherwise the long term stability may be poor as the strain in the wire relaxes with time. In extreme cases with very fine wire windings lack of attention to the required strain-free construction may cause the winding to become open circuit.

The wire used should also be of a material having a low temperature coefficient of resistance, which is particularly important if the multiplier is intended for use over a wide ambient temperature range or if the temperature rise due to self-heating is likely to be high.

If wire wound resistors are intended for use on a.c., unless they are of the card type, some form of non-inductive winding such as bifilar is necessary, and if the frequency range is more than a few hundred hertz then careful design consideration must be given to achieving a suitably low time constant for the finished multiplier.

Wire wound resistors do not normally exhibit any significant voltage coefficient so this effect may be ignored, but as with all types of multiplier care must be taken with the design of connections and insulators to ensure that the resistance accuracy is not degraded by corona or leakage effects under working conditions.

With film resistors account must be taken of the effect of voltage coefficient. In modern resistors this effect is normally quite small of the order of 0.001% or less per volt, but if the highest accuracy is required due allowance must be made in order to obtain the required resistance at working voltage.

If, for insulation purposes, or even for purely mechanical reasons series resistance multipliers are ' potted ' considerable care must be exercised in the choice of potting material and the method of application. The material itself must obviously have extremely good electrical characteristics, such as high insulation resistance, anti-tracking properties, and if for use with a.c. multipliers, a low dielectric constant is desirable. It has been found by experience that certain of the two pack epoxy materials should be avoided, owing to the high exothermic temperature rise which occurs during curing of the compound. This in certain instances can be offset by the inclusion of a fairly high proportion of inactive fillers such as marble flour etc., although these are inclined to make the mix rather stodgy and difficult to pour. A material which has a slightly resilient nature when fully cured has also been found desirable as there is less tendency to introduce strain into the resistors being potted. A material having low shrinkage during cure should be chosen for the same reason. For potted multipliers vacuum degassing immediately following pouring has been found advantageous as this process eliminates voids in the potting compound which, if allowed to remain, may cause flashover and contribute to general electrical noise and ionization currents.

For high voltage multipliers considerable care is also needed in the design of the stand-off insulator which is usually necessary for the ' hot ' connection. As long a tracking path as possible should be allowed and if the multiplier is likely to be used in dirty or dusty atmospheres, then an actual breakdown voltage for the insulator of several times the actual working voltage is highly desirable as the performance in use is liable to deteriorate considerably as dirt and dust collect on the surface during use. The insulator should be made of a good quality glazed ceramic or one of the plastics such as polythene which have good anti-tracking properties.

For added safety in use it is normal to make high voltage multipliers as three-terminal devices as shown in the circuit of fig. 8.1. This

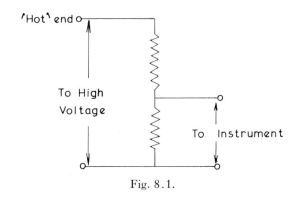

Fig. 8.1.

ensures that if the instrument becomes open-circuited for any reason, the voltage on the instrument terminals does not rise to the high voltage on the 'hot' end of the multiplier, which could easily be lethal. The value for R_2 which shunts the instrument must obviously be chosen to be sufficiently high compared with the instrument resistance so as not to have any significant effect on the accuracy of the measurement. Conversely the value of R_2 should also be as low as possible so that the voltage to which the junction between R_1 and R_2 can rise if the instrument circuit is opened, is kept as low as possible. These two conflicting requirements obviously can only be met by a compromise value. Figure 8.2 is a picture of a 20 kV multiplier made by Sangamo Weston Ltd. at 10 000 ohms per volt, so that R_1 is 200 megohms. R_2 is 500 000 ohms, which means that providing the instrument resistance is not more than 500 ohms the shunting effect of R_2 will not introduce an error greater than 0·1%, whilst if the instrument circuit is opened the junction of R_1 and R_2 rises only to about 50 volts.

Shunts

If it is required to measure direct currents of more than a few mA by means of a moving coil instrument, then a suitable shunt is necessary. A shunt is just a low resistance which is connected in

Fig. 8.2.

parallel (or shunt) with the instrument so that the major proportion of the current being measured passes through the shunt. The instrument used for this type of arrangement is usually a millivoltmeter, 50 or 75 mV full scale deflection being commonly used, and the resistance of the shunt being such as to provide the appropriate potential difference across its terminals when the specified current is passing through it.

For industrial grade instruments the shunt is often permanently connected across the instrument terminals for currents up to about 50 amperes, but for higher currents, or greater precision the shunt is made as a separate four-terminal device. Two terminals are provided of a suitable size, according to the current rating, which carry the main current and two potential terminals for connection to the millivoltmeter.

Many different constructions have been used for shunts, each with a claimed advantage for a particular installation or range or for ease of adjustment etc., but the most commonly used form consists of two copper blocks between which are soldered one or more strips of suitable resistance material such as constantan, manganin or minalpha. The size of the copper blocks and the number and size of the resistance strips is obviously dependent on the current range and required millivolt drop but this general construction has been used from about 10 amps up to several thousand amps quite satisfactorily. A typical example of a 500 ampere shunt is shown in fig. 8.3. This shunt was made by Sangamo Weston Ltd. and has a drop of 50 mV and is offered at an accuracy of $\mp 1\%$.

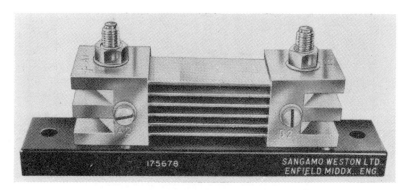

Fig. 8.3.

The detail design of shunts of this type involves consideration of a number of factors, particularly those concerned with the calculation of the permissible temperature rise under working conditions, since the wattage dissipation involved, especially with the higher current ranges, is quite high. For example, a 1000 ampere shunt having a drop of 50 mV dissipates 50 watts, without allowing for the inevitable small additional dissipation in the copper blocks and terminations. The design, therefore, must make adequate provision for the resulting heat to be dissipated without an excessive temperature rise.

It has been shown by various experiments that the heat loss by convection from a flat plate is very dependent on the position and form of the plate and is also affected by the finish of the plate. The percentage of the total heat loss by conduction may also vary from 60% to 90% dependent on the same factors, therefore, it is very difficult to settle on a value for the emissivity which may be used in a rigorous calculation of a shunt design for a particular temperature rise.

A general figure which has been found to give satisfactory results is to allow 1 watt per square inch of total surface area of plates. This has been found in most cases to give a temperature rise at the centre of about 50°C, but this figure should only be used as a guide and considerable experience is needed to judge how far it is to be relied on in any particular application.

As mentioned earlier the general construction described is suitable for currents up to several thousand amperes and single unit shunts have been made up to at least 10 000 amperes. Above this value, however, the problems of ensuring adequate connections and the difficulties of testing the shunt under working conditions make this construction unsuitable and the most usual method to overcome the problem is to use a number of shunts in parallel. The potential terminals are also connected in parallel by means of leads of equal resistance and the junctions taken to the terminals of the millivoltmeter. For example, a current of 20 000 amperes can be measured by using four shunts each of 5000 ampere range. A diagram of the connection of such an arrangement is shown in fig. 8.4.

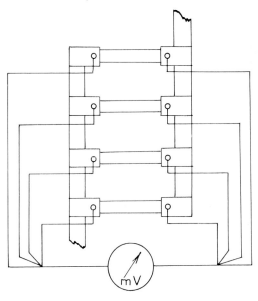

Fig. 8.4.

Rectifiers

The usefulness of the moving coil instrument can be considerably extended by the inclusion of a rectifier in its circuit thereby making it suitable for the measurement of alternating currents. A rectifier is essentially a non-linear device which possesses the property of passing a current more easily in one direction than the other. This property is exhibited by a number of devices and materials ranging from the thermionic valve to the more recently introduced semiconducting diodes made from silicon and germanium. Most of these devices have at some time been used in conjunction with a moving coil instrument but the most commonly encountered are copper–oxide, silicon or germanium, for normal rectifier instruments. In each instance the basic principle is the same and the particular choice is decided by factors such as frequency, ambient temperature, etc.

The simplest possible circuit arrangement would be that shown in fig. 8.5 consisting of a single junction in series with the moving coil.

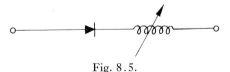

Fig. 8.5.

Such an arrangement is, however, rarely used, partly because the full peak inverse voltage of the system is impressed across the junction during the non-conducting half-cycle and partly because only alternate half-cycles cause any current flow through the instrument. A slight improvement on this arrangement is shown in fig. 8.6, which overcomes

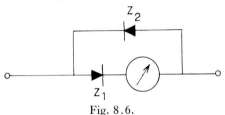

Fig. 8.6.

the first objection, since the peak inverse voltage for each rectifier is now only the forward volt drop of the other (plus for Z_2 the volt drop across the instrument).

The more usual arrangement employing a full wave bridge rectifier is shown in fig. 8.7. This circuit overcomes both the objections to the

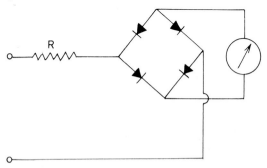

Fig. 8.7.

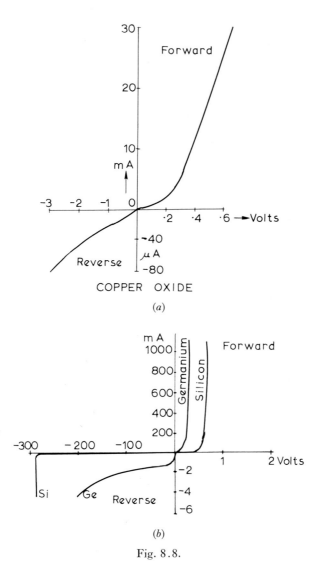

Fig. 8.8.

previously described half wave circuits. Resistor R turns the arrangement into a voltmeter and is omitted when current measurement is required. It will be noticed that for each of the half-waves of an alternating current two of the four diodes are working in the forward of conducting direction and two in the reverse or non-conducting direction. Whilst the connections of the bridge are such that the current in the instrument is undirectional and consists of one pulse for each half-wave of the input.

Now if the diodes were perfect each pulse would consist of a half-wave of current whose instantaneous value at any time was directly proportional to the instantaneous value of the voltage impressed across the circuit. Since, however, no diodes are perfect there are two effects which have to be considered at low frequencies. First, the diode reverse resistance is not infinite, therefore a small reverse current flows, thus reducing the magnitude of the forward current. Secondly, the forward resistance is a function of the instantaneous current and as it is varying throughout the half-cycle will cause some difference between the voltage and current waveforms. This second effect is most noticeable on low range voltmeters where the change in resistance throughout the half-cycle is a significant proportion of the total circuit resistance. As the value of R is increased this effect is reduced. This non-linear relationship between current and voltage also accounts for the non-linear scales of low range voltmeters, the effect becoming more pronounced the lower the range. Typical curves of the current/resistance relationship for copper oxide, germanium and silicon rectifiers are shown in fig. 8.8. The overall effect is also made more complicated as frequency increases since the parameters are not entirely resistive, but contain a capacitive component which can give rise to errors of several per cent at 10 000 Hz on a copper–oxide rectifier.

CHAPTER 9

the special problems of aircraft instruments

THE electrical measuring instruments used in aircraft are largely moving coil of either the normal linear scale type measuring current of voltage or ratiometers used in conjunction with various transducers for the measurement of temperature, pressure, position, etc. In addition a few iron-covered dynamometers have been used for power measurement. Moving iron, electrostatic and induction instruments are rarely if ever employed.

The special problems associated with the design and manufacture of aircraft electrical instruments are always concerned with one or more of the following factors:

(1) small available space, (2) adverse reading conditions combined with high accuracy requirements, (3) adverse environmental conditions, (4) high reliability combined with minimum weight.

The influence of each of these factors has had a particular effect on the design trend, resulting in styles of instrument which today bear little resemblance to the industrial instruments from which they were originally derived, although industrial instruments have undoubtedly benefited from some of the enormous amount of development which has been necessary to satisfy the rigorous demands of the modern aircraft industry for electrical instruments.

It is interesting to examine in a little more detail the design trends which have resulted from the factors set out above.

1. *Small available space*

This design limitation results from the understandable aircraft requirements of minimum weight coupled with the necessity to present a large number of indications on a cockpit panel of very limited area. These considerations in particular have led to instruments of smaller and smaller frontal area and to a lesser extent a reduction in overall instrument volume. In order to make greater use of the scale area available, development has followed two distinct lines of approach, namely circular scale presentations and multimovement presentations. Typical examples of these two approaches are shown in fig. 9.1. It should be noted that each of these presentations is housed in a case of only 2 in. nominal diameter. An interesting comparison between these two examples is the total scale length presented in the 2 in. diameter case by these two methods. The circular scale example has a scale length of

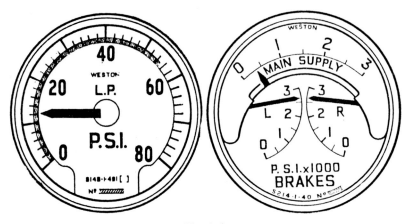

Fig. 9.1.

3·3 in. whilst the three-movement example has individual scale lengths of 1·4 in. and 0·8 in. totalling 3·0 in.

2. *Adverse reading conditions combined with high accuracy*

The adverse reading conditions which often prevail in aircraft are partly associated with item 1 in that owing to the limited panel space available many instruments cannot be duplicated at the pilot and co-pilot positions, so it is necessary to place instruments so that they are always read at an angle which is far removed from the desirable face-on condition. The reading conditions are often further degraded by the very variable lighting conditions which prevail during night flying, bright sunlight, etc. Undesirable reflections from the instrument windows may also add to the difficulties. These problems have all been tackled by one or more methods, but it would be futile to suggest that the solutions so far discovered are ideal. The problem of reading at an angle has been overcome to a reasonable extent by the use of ' platform ' scales described in an earlier chapter. Many aircraft instruments are now ' integrally ' lit making them less dependent on the external lighting conditions for adequate visibility. By ' integral ' lighting is meant the system by which scale and pointer illumination lamps are built into the instrument. Two basic methods are in common use, known respectively as ' back lighting ' and ' wedge lighting '. Figure 9.2 shows diagrammatic sections through typical examples of these two types. The choice of which type of lighting should be used in any particular installation is largely a matter of personal preference on the part of the designer, but the wedge lighting has a small advantage from the manufacturer's point of view, in that it is more versatile. Once a system has been designed for a particular case size or style, then wedge lighting is almost entirely independent of the type of mechanism which is put in the case,

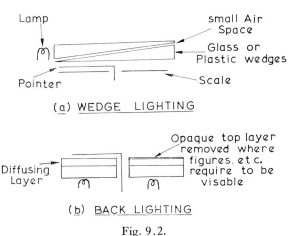

Fig. 9.2.

whereas, with back lighting the siting of lamps and light guides may have to be individually tailored for each type of mechanism. For example, the two different scale presentations shown in fig. 9.1 can be 'wedge' lit by the same case design, whereas the back lighting design for these same two examples would almost certainly have to be entirely different. The small lamps used for integral lighting are special rugged filament, low voltage, lamps under-run to ensure long lamp life. The usual systems comprise 6 volt lamps run at a maximum of 5 volts with single or group voltage controls available to the pilot by which he can adjust the level of illumination to his own particular preference. Many integrally lit and other instruments incorporate windows or lenses which have been specially 'bloomed' to reduce unwanted reflections.

3. *Adverse environmental conditions*

The range and variety of environmental conditions under which aircraft instruments are required to operate is truly enormous. Partly this environment is dependent on the aircraft itself which imposes vibration, acceleration and altitude and partly by the extremes of temperature, humidity, etc., which the whole aircraft can experience in its rapid journeys from one part of the world to another. Yet another set of conditions may be encountered during emergency or other non-standard procedures such as contamination by fuels, hydraulic fluids, etc., or shock conditions due to heavy landings, minor crashes, etc. A great deal of information regarding the likely environments and suitable laboratory test procedures to ensure satisfactory performance in service is given in B.S. 2 G100 for civil aircraft and the equivalent military aircraft document AVP 24.

To give only one example of the conditions which must be catered for, cockpit instruments may be required to function over a temperature range of $-40°C$ to $+70°C$ and not be deranged by subjection to extreme temperatures of $-55°C$ to $+90°C$.

A great deal of design and test work is obviously necessary to produce instruments which will function under such conditions and has led to the increasing use of sealed instruments, ruggedized movement mountings, modified bearing systems, special springs, tropical finishes, sealed connectors, etc. to name but a few of the more important features.

4. *High reliability combined with minimum weight*

To a large extent the requirement of high reliability is coupled with the previous factors requiring satisfactory functioning under adverse conditions but it also is a prime necessity from the safety aspect. Modern civil aircraft may be actually flying for upwards of 3000 hours per year, so it is obviously necessary that aircraft equipment of all types, including instruments, should be designed for long trouble-free life without frequent overhaul, since the safety of the aircraft and its passengers may be vitally dependent upon its correct functioning. Even where instruments are not vital to flight safety malfunction can represent expensive and annoying delays.

These potential difficulties explain the increasing pressure placed by both airframe and engine manufacturers alike on equipment manufacturers to design and produce items in which reliability has been given the greatest possible consideration. Reliability is written into many commercial contracts and guarantees are required that predicted M.T.B.F.'s (mean time between failures) will be met. Failure to do so may involve considerable financial penalties. When these various factors are combined with the necessity to keep weight to a minimum it is obvious that some quite sophisticated design and experimental techniques need to be employed in order to meet the apparently conflicting requirements.

Design approval

Most companies engaged in the manufacture of instruments and other equipment for aircraft hold certificates of design approval from the Ministry of Technology and/or the Air Registration Board. By this is meant that one or both of these authorities have vetted the company's design capabilities from both a personnel and equipment point of view and have certified their competence to design and test a specified range of aircraft products.

This arrangement is usually linked with a separate inspection approval signifying that the company has adequate inspection personnel, equipment and procedures to ensure that manufacture and test complies with the relevant specifications and drawings.

To illustrate by way of example the various stages required to be gone through from inception to final delivery, let us imagine that a requirement is established for a completely new design of aircraft instrument. Let us also imagine that the requirement is not specific to one customer, but has arisen due to market survey information that there would be a demand for a new miniature instrument of, let us say, 1 in. diameter. The market survey would also include information on the likely range requirements, annual quantity and a provisional price.

If, on the basis of this information, the company management decided that the potential was worth investigating, all the available information would be passed to a designer for a feasibility study to be carried out. This would permit expenditure up to a certain amount over a limited period, say 1 or 2 months. During this time it would be the designer's task to examine possible constructions, bearing in mind the ultimate necessity to meet general specifications such as B.S. 2 G100 and AVP 24 as previously mentioned. He would also prepare sufficient information to enable a preliminary estimate to be made of tooling and manufacturing costs and also a development programme showing likely development time and cost. He would probably also produce a prototype model illustrating the basic construction. If all this additional information was favourable, instructions would then be given by management to proceed to full development and production.

Now the designer's task would commence in earnest. If the proposed construction was similar in any respect to existing designs considerable information would be available regarding their performance under varying environmental conditions such as vibration, temperature, etc., and the designer could assess the relevance of such information to his new design. Where the new design differed from current practice it would be his responsibility to have assemblies or sub-assemblies made so that the earliest possible assessment could be made by actual test of the performance and modifications made as necessary. While design was proceeding he would also be required to ensure by consultation with tooling, production and other experts that his design was compatible with low cost, ease of production and reliability. For this reason any proposed new materials or manufacturing techniques would come under particularly close scrutiny. Liaison would also be maintained throughout development with the sales and marketing function to ensure that the emergent design and the customer environment were still compatible.

As soon as a first fully functioning prototype was available, environmental testing such as a vibration resonance search, temperature range and coefficient tests would be carried out, whilst detailed material specifications for purchasing were being prepared together with manufacturing specifications and assessment of production test equipment requirements.

By this time the actual design would be nearly finalized, modifications having been incorporated as necessary according to the results of tests etc., and one or more models would be available for test and assessment, possibly by potential customers, who had shown a particular interest. It will be seen that by now the project has spread from the designer's original conception to embrace a number of other people who perform various functions in parallel as the design proceeds. As soon as possible all the differing requirements are reconciled, the design finalized and production drawings are issued for tooling to begin, although certain preliminary tooling may have been necessary already to confirm new methods or processes.

Whilst tooling proceeds a complete programme of type test and reliability assessment is undertaken to confirm that the performance agrees with the specification requirements. Inspection and quality control functions are also busy preparing the necessary gauges and inspection and quality control procedures. Materials and components have been provisioned and the production departments have been arranging for suitable space, labour and machinery to be available when required.

The first production made instruments are made available to the designer who arranges for such tests as are deemed necessary to be carried out and any final alterations to be incorporated. When the designer is satisfied and any queries or difficulties encountered in the manufacture of these first production samples have been resolved, the product is released for production.

During the latter stages of tooling, etc., the sales department will have been fully informed of progress, and will probably have accepted initial customers' orders so that production can commence as soon as the final release is forthcoming.

Production and inspection departments will have already prepared the various procedures which are necessary to ensure that the history of all important piece parts which go to make up an instrument can be traced back to raw material batches and the appropriate E.I.D. or A.R.B. release notes called up by the original purchase specifications.

Once production has been satisfactorily established the product is largely left to the care of the production, inspection and quality control departments, whose responsibility it is to ensure that the original specifications laid down continue to be met. This is monitored by the various inspection stages throughout manufacture and also by random sampling of the finished product.

The reliability assessment and test programme which were undertaken earlier will have resulted in a predicted reliability quoted as mean time between failures (M.T.B.F.) and it now becomes the job of the product support group to ensure that all information fed back from the field is collated and assessed by the reliability engineer, who in consultation with the original designer and other interested parties will instigate any design or procedure modifications which are deemed necessary as a result of that information. This is a continuous process which extends for the whole of the time that the product is manufactured and the overall aim is a continuous improvement in the reliability.

Apart from the original type tests carried out to confirm performance under various environmental conditions, it is commonly laid down that complete type tests should be repeated every 2 or 3 years. This is to ensure that as a result of minor changes in design or procedure made over a period the original performance is not impaired.

It will be seen from the brief description above that the manufacture of aircraft instruments involves a degree of control during all stages of design and manufacture, which could not normally be justified for industrial products. This procedure is obviously reflected in the price of the product, but the benefits which accrue due to added safety and reliability are deemed to be well worth while.

CHAPTER 10
transducers

FOR the purposes of this chapter, transducers are defined as devices which when acted upon by any particular physical or chemical parameter provide an output suitable, sometimes in conjunction with additional circuitry, for energizing an electrical indicating instrument, so that it may be calibrated in terms of a non-electrical quantity. It should be appreciated that this is a narrow and specific definition of transducers which excludes a high proportion of the types available meeting a broader definition.

Transducers even as defined here, cover a vast and bewildering range and only examples of the simpler and more common types are described in order to illustrate the basic principles involved. Some of the most commonly encountered transducers are used for the measurement of temperature, pressure, position, speed and torque, and examples of all these are given.

Temperature transducers

Temperature transducers for use with electrical indicating instruments fall broadly into two categories:

(*a*) *resistive types and* (*b*) *thermocouples*

The resistive types rely on the property of certain materials which change their resistance in a predictable manner according to their temperature. One of the most widely used examples of this type is the platinum resistance thermometer, which in various forms has been made suitable for the accurate measurement of temperature over a range approaching absolute zero up to 1000°C or more.

The platinum thermometer relies on the property of high purity platinum that its resistance follows a predictable law of the form:

$$R_T = R_0 (1 + \alpha t - \beta t^3),$$

where R_T = resistance at any temperature T, R_0 = resistance at a temperature of 0°C, and α and β are constants depending on the purity and mechanical condition of the platinum.

For commercial high purity platinum α and β are of the order of 0·00392 and 0·00000058 respectively.

The platinum is normally used in the form of wire wound on to an insulating former with suitable leads and mechanical protection. A

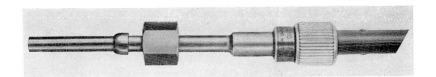

Fig. 10.1.

typical example made by Sangamo Weston Ltd. is shown in fig. 10.1. In this instance the platinum wire has a diameter of 0.0016 in. and is wound as a coiled coil on to an anodized aluminium former. The platinum winding is adjusted to a resistance of 130 ohms at 0°C and is protected by an anodized aluminium cap which is slid over the former. The whole assembly is then slid into a $\frac{1}{4}$ in. outside diameter stainless steel sheath, connection being made either by a miniature plug and socket or by flying leads. The thermometer is normally air filled but where response time is important it can be supplied filled with hydrogen. The response time in water for the air filled version is approximately 7 sec whereas in the hydrogen filled version this is reduced to approximately 3 or 4 sec.

Platinum resistance thermometers are often referred to by what is known as their 'fundamental interval'. This is merely $(R_{100} - R_0)$ and for a thermometer having a resistance at 0°C of 100 ohms is approximately 39.5 ohms and for one where $R_0 = 130$ ohms is approximately 50 ohms.

For use with indicating instruments it is usual to employ thermometers with high resistance values, as this makes lead resistance less important, as well as being compatible with instrument circuit resistances, but for the greatest accuracy under laboratory conditions thermometers with resistances of down to 10 ohms or less can be employed. These low resistance thermometers can generally be made to a better stability, as the platinum wire can be 0·008 in. diameter or even larger and consequently less susceptible to surface contamination.

For similar reasons the low resistance thermometers are normally used for the measurement of temperatures above about 600°C and various special three and four lead circuits are used to reduce errors due to lead resistance. Platinum resistance thermometers are made in a variety of shapes, sizes and resistances for different applications. With extreme care during manufacture to ensure a strain free, uncontaminated platinum element the resistance figures are so reproducible that the thermometer is used as the standard for interpolation of the International Temperature Scale between the fixed points provided by the boiling point of oxygen of −182·97°C up to 660°C which is the melting point of aluminium.

Other materials are used for resistance thermometers such as nickel, copper and semiconductors. They are in general not as stable as platinum but have the advantage of being cheaper where accuracy and repeatability are not the prime considerations and in the case of the semiconductor types a larger temperature coefficient of resistance can be obtained which allows the use of simpler measuring equipment.

Thermocouples

Thermocouples rely for their function on the basic principle that when the junction of two dissimilar conductors (usually metals) is heated an e.m.f. is generated which is a function of the temperature difference between the hot and cold junctions.

An enormous variety of materials are used for thermocouples some of the most popular being copper versus constantan, iron versus constantan, nickel–chromium versus nickel–aluminium and platinum versus a variety of platinum alloys. The reasons for choosing any particular combination are very complex involving consideration of some or all of the following: cost, temperature range, millivolt output, stability, corrosive or erosive environment, contamination and electrical resistance. Suffice it to say that copper/constantan is used for the lower temperature range up to about 300°C to 400°C, nickel–chromium/nickel–aluminium for the middle temperature range up to say 1000°C and the platinum types above this up to perhaps 1800°C.

The output of a thermocouple is dependent on the materials used in the junction and the temperature difference between the hot and cold junctions. It is impossible to connect a thermocouple to any form of measuring apparatus without forming at least one other junction (the cold junction). The generated e.m.f. of a single hot junction is never very high, only a few tens of millivolts at most, so that indicating instruments to go with them are usually very sensitive and consequently fairly delicate.

Pressure transducers

Pressure transducers are made in a wide variety of ranges from a few inches water gauge up to thousands of pounds per square inch for full scale output. Most types rely on various forms of bellows, diaphragms or bourden tubes to convert the applied pressures into mechanical movement. For use with electrical indicating instruments this movement has to be converted into some form of electrical signal compatible with an indicator. This may be achieved by arranging for the mechanical movement to vary an inductance, a capacitance, a resistance or the magnetic coupling in a transformer. A great many ingenious variations and combinations of these principles have been employed and each may have its merits in particular circumstances. The types involving a resistance change are some of the simplest and are usually suitable for d.c. energization. They generally rely for operation either on the change of resistance in a strain gauge when a strain is applied or on some

Fig. 10.2.

form of sliding contact moving over a resistance track. An example of the latter type as manufactured by Sangamo Weston Ltd. is illustrated in fig. 10.2. This device is available in various pressure ranges from about 0–8 p.s.i. to 0–300 p.s.i. It incorporates a bellows and spring arrangement which gives a mechanical movement of approximately 0·1 in. for full scale pressure. A similar device based on a serpentine bourden tube is also manufactured by the same company and covers higher ranges up to 6000 p.s.i. It also has a mechanical movement of approximately 0·1 in. In both types this movement causes the sliding contact of a potentiometer to traverse approximately 0·4 in. along the potentiometer. This magnification of the original movement is achieved by the relative angles of the contact and winding.

It should be remembered that all pressure gauges are inherently differential devices, that is, they measure the difference in pressure between the two sides of the pressure sensitive component, whether this is a capsule, a diaphragm, a bellows or a bourden tube. If one side is vented to atmosphere then the indications are in p.s.i. above or below atmospheric and are termed p.s.i. gauge. If one side is evacuated and sealed then the indications will be in p.s.i. absolute. The third possible variation is when both sides of the diaphragm etc. are pressurized from an external source. In this case the indications will be of differential pressure. By the use of various materials and constructions transducers are made for the measurement of gas and liquid pressures including highly corrosive media such as are encountered in certain industrial and aircraft applications.

Position transducers

Position transducers are also made in considerable variety and range from sophisticated electronic devices for the measurement of exceedingly small movements down to a few millionths of an inch up to devices for measuring a movement of many feet. Both rotary and linear movements can be measured and as with pressure transducers there are a number of ways in which this movement is converted to an electrical signal. Again one of the simplest ways of conversion is by means of a variation in resistance due to the motion of a sliding contact on a resistance winding. An example of a rotary device of this type is shown diagrammatically in fig. 10.3. This example is intended for use with a d.c. ratiometer

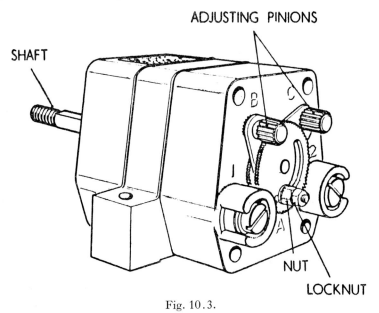

Fig. 10.3.

indicator and incorporates an ingenious arrangement of fixed and sliding contacts which can be adjusted to make one standard indicator read full scale deflection for any angular movement between approximately 30 and 300 angular degrees. The circuit for this device with its indicator is shown in fig. 10.4. The two adjustable fixed contacts A and B can be adjusted and locked externally after the transducer has been installed. This device has been used extensively for the position indication of aircraft control surfaces such as rudders, ailerons, flaps, etc. as well as for industrial purposes. Potentiometric devices are also used in a similar manner for the measurement of linear movement.

Speed transducers

Speed transducers are normally termed tachometer generators. Most measurements of speed are rotary speeds in revolutions per minute but where the measurement required is of linear speed this is

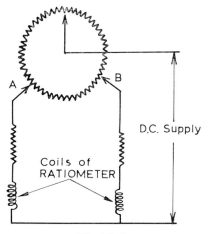

Fig. 10.4.

usually converted to a rotary speed by gearing etc. before measurement.

Tachometer generators are usually some form of electrical alternator or d.c. generator frequently employing permanent magnet rotors or stators. The parameter which is actually measured is either the voltage or frequency output of the generator, the frequency measurement being capable of the greater accuracy since the frequency is a design feature, unlike the voltage which is dependent on adjustment during manufacture.

One of the simpler and more versatile devices used in speed measurement is what is known as a speed probe. This is used for rotary measurement by detecting the passing of the teeth on a suitable ferrous gearwheel or cam. The principle is illustrated in fig. 10.5 and the

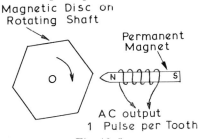

Fig. 10.5.

output frequency is a direct function of the speed of rotation of the gear. It may be considered as a variable reluctance generator with each tooth as it passes the tip of the probe altering the reluctance of the magnetic circuit embraced by the coil. This device is commonly used with an electronic counter to indicate the speed but by suitable amplification of the probe output the signal can be made suitable for energizing a frequency meter, calibrated in revolutions per minute etc.

Torque transducers

The measurement of transmitted torque is an important measurement in various high speed and/or high power shafts such as are used in helicopters and ships. A number of methods have been employed including optical, strain gauge and magnetostriction types. The principle employed is normally the detection of the twist angle occurring along the length of a specially designed shaft which is part of the power transmission. One of the practical difficulties in strain gauge and magnetostriction types is in transferring the signal from the rotating shaft to the stationary measuring circuit. This can be achieved by slip rings and brushes but these tend to be unreliable at the low voltages normally employed. Special rotary transformers have been developed to overcome this problem but they also are rather complicated devices and prone to various forms of interference.

An old basic method which has been the subject of considerable recent development is what is known as a phase displacement torquemeter. This consists essentially of an arrangement of two alternators one at each end of the specially designed transmitting shaft. By arranging that the outputs of the two generators are in phase when the torque being transmitted is zero then providing the angle of twist of the shaft is known for a particular value of transmitted torque then a measurement of the phase relationship between the two alternator outputs can be translated into torque. Considerable magnification of the original angular twist in the shaft can be made by the use of multipole alternators as the following example will show:

If the angular twist between the two ends of the shaft for maximum permitted torque is, say, one angular degree then if an alternator having fifty poles is used this will cause a change in phase between the two alternators of fifty electrical degrees. By the use of suitable electronic circuits this angle of fifty electrical degrees can be measured to a high accuracy and thus the torque indicated. A considerable degree of mechanical precision is required in the manufacture of the shaft and the alternators but the method has much to recommend it on the grounds of its sound basic principles.

There are, of course, many more types and principles for transducers than have been mentioned, but it is hoped that the above descriptions of examples of the more common and simpler types will serve to illustrate some of the ways in which electrical indicating instruments can be used for the measurement of non-electrical quantities.

CHAPTER 11
what of the future?

OVER the past decade or so the simple electrical measuring instrument has seen many of the roles at either end of its range of capabilities taken over by other devices. At its simplest, when used as a comparatively rough indicator, in the interests of economy and space saving it has been replaced by a lamp. The modern mass produced motor car contains several examples of this. Gone are the ammeters and the oil pressure gauges of yesteryear to be replaced by coloured lights which should light up when things go wrong. It is interesting to note that in spite of all the arguments about how much more positive an indication one gets from a lamp most of the more expensive cars still retain the so called ' old fashioned ' instruments for these functions.

At the other end of the scale where the demand is for greater and greater accuracy the role of the indicating instrument is rapidly being taken over by electronic devices based on the digital voltmeter, which are capable of an accuracy in general at least one order better than the best indicating instruments. Digital instruments also have the added advantage of requiring little skill in reading and usually have the additional facility of providing a print-out for permanent record where this is required. This trend to replace indicating instruments by digital devices will undoubtedly continue, limited largely by cost and size, both of which are being continually reduced.

It would seem from this brief analysis that the future outlook for electrical indicating instruments is very bleak. Fortunately this does not appear to be so. First, at the same time as it is being replaced by other devices, the overall demand for measuring apparatus is increasing to such an extent that the number of indicating instruments required shows no significant decline. Secondly the indicating instrument possesses one feature which no alternative device has so far been able to replace. This is its ability to indicate trends. So long as the human eye is part of the control loop for any machine or process the ability of an indicating instrument to show the trend of any measurement is likely to remain a very important attribute.

Many attempts have been made to design an indicating instrument which could rival the digital devices in accuracy or readability or both but these have met with only very limited success, being mostly complicated and unreliable devices, relying on complex optical or mechanical amplification to obtain the required effective scale lengths. Perhaps the effective combination of the best features of indicating instruments and digital devices lies in some invention not seen so far, but it is a very salutary thought for the instrument designer that no basically new principle for electrical indicating instruments has been discovered in the past 80 or 90 years.

Another pressure which is being brought to bear on instrument designers is reduction in size. This pressure is brought about largely by the enormous advances in the miniaturization of electronic equipment with the consequent reduction in panel area, leaving as a result less space for the mounting of an instrument. This has called for detailed consideration by the instrument designer of ways of making the most efficient use of the panel area available. The problem has been tackled in several different ways such as the increased use of circular scale and edgewise mechanisms, dual purpose instruments which replace two or more single instruments and the general slimming down of instrument cases.

Circular scale instruments are probably the most efficient style for minimum use of panel space, closely followed by edgewise designs, particularly where a shallow mechanism is employed and the instrument width reduced accordingly. Edgewise types also have the added advantage with certain designs that they can be stacked closely together. Both circular scale and edgewise designs have, however, the disadvantage of occupying a relatively large amount of 'behind the panel' space.

Dual purpose instruments such as the volt-ammeters used on many stabilized power units are popular space-savers where simultaneous readings of the two parameters are not required.

These and other similar trends will undoubtedly continue in the future and as further miniaturization of electronics proceeds the ingenuity of designers of indicating instruments will inevitably be stretched to keep pace with it.

It is impossible to forecast the ultimate future of indicating instruments as we know them today, but it is a fairly safe assumption that for the next decade or so, in the majority of applications they will at least be very similar to those with which we are familiar. It is interesting to speculate, however, on possible developments which may emerge. Solid state devices seem the most likely area for any major breakthrough to occur. Relatively crude devices have been demonstrated in which a number of magnetic cores or diodes have been arranged to be progressively activated by increasing increments of current or voltage and by suitable circuitry to energize neons, electrofluorescent lamps or semiconductor light emitters. So far these have been bulky, expensive and of indifferent accuracy, but if sufficient development effort were applied it is conceivable to imagine a completely solid state device consisting of say, 100 discrete light emitting spots arranged so that they light up progressively as the applied currents or voltage increases. These could be arranged against a scale so that instead of a pointer a light spot or band performed the indicating function. The perfection of such a device at a reasonable cost would without doubt stimulate the development of circuitry to allow its use for measuring many, if not all, the parameters for which present electrical indicating instruments are used.

To sum up, therefore, although the appearance or even the principle on which future indicating instruments function may well change, it is difficult to visualize a state of technology where some form of instrument, presenting analogue information, does not form an important part of the tools of progress.

INDEX

Ampere 5
Anodizing 72
Antiparallax Devices 78
Ayrton and Perry 6

Bakelite 67
Balance 10
Brass 60

Cellulose 67
Coulomb 5

Damping 43–46
Die Castings 61
Dynamometer 22–35
  Ammeter 27
  Voltmeter 24
  Wattmeter 29–35
  L.P.F. Wattmeter 30

Electrostatic instrument 7, 35, 42

Faraday 6
Finishes 72
Frequency errors 19–20

Glass 62

Induction instrument 36
Integral lighting 98

Kelvin 5, 6, 78

Magnets 5, 11, 13, 64–66
Magnet raising 85
Magnetic impurities 57
Methylmethacrylate 68
Mild steel 60
Moving coil 9–17, 39, 84
Moving iron 17–22, 39–41

Multipliers 88

Oersted 5

Perspex 68
Pivots 62, 84
Plastics 66–69
Plasticizer 70
Plating 72
Pointers 77
Polyamides 68
Polycarbonate 69
Polystyrene 69
Position transducer 107
Pressure transducer 106

Ratiometer 15–17
Rectifiers 93–96
  Copper oxide 94, 95
  Silicon 94, 95
  Germanium 94, 95
Resistance thermometer 104
Resistors 89

Scales 75
Schweigger 5
Self-heating error 32
Shunts 90–93
Springs 62

Tachometers 109
Taut band 47–50
Temperature errors 50–52
Thermocouples 104–106
Torque to weight 52
Torque transducer 110

Weston 1, 6
Wire 54–59

THE WYKEHAM SCIENCE SERIES
for schools and universities

1. *Elementary Science of Metals* J. W. MARTIN and R. A. HULL
   (S.B. No. 85109 010 9) 20s.—£1.00 net *in U.K. only*

2. *Neutron Physics* G. E. BACON and G. R. NOAKES
   (S.B. No. 85109 020 6) 20s.—£1.00 net *in U.K. only*

3. *Essentials of Meteorology* D. H. McINTOSH,
   (S.B. No. 85109 040 0) A. S. THOM and V. T. SAUNDERS
   20s.—£1.00 net *in U.K. only*

4. *Nuclear Fusion* H. R. HULME and A. McB. COLLIEU
   (S.B. No. 85109 050 8) 20s.—£1.00 net *in U.K. only*

5. *Water Waves* N. F. BARBER and G. GHEY
   (S.B. No. 85109 060 5) 20s.—£1.00 net *in U.K. only*

6. *Gravity and the Earth* A. H. COOK and V. T. SAUNDERS
   (S.B. No. 85109 070 2) 20s.—£1.00 net *in U.K. only*

7. *Relativity and High Energy Physics* W. G. V. ROSSER
   (S.B. No. 85109 080 X) and R. K. McCULLOCH
   20s.—£1.00 net *in U.K. only*

8. *The Method of Science* R. HARRÉ and D. EASTWOOD
   (ISBN 0 85109 090 7) 25s.—£1.25 net *in U.K. only*

9. *Introduction to Polymer Science* L. R. G. TRELOAR
   (ISBN 0 85109 100 8) and W. F. ARCHENHOLD
   30s.—£1.5 net *in U.K. only*

10. *The Stars: their structure and evolution* R. J. TAYLER
    (ISBN 0 85109 110 5) and A. S. EVEREST
    30s.—£1.5 net *in U.K. only*

11. *Superconductivity* A. W. D. TAYLOR and G. R. NOAKES
    (ISBN 0 85109 120 2) 25s.—£1.25 net *in U.K. only*

THE WYKEHAM TECHNOLOGICAL SERIES
for universities and institutes of technology

1. *Frequency Conversion* J. THOMSON,
   (S.B. No. 85109 030 3) W. E. TURK and M. BEESLEY

2. *The Art and Science of Electrical Measuring Instruments*
   (ISBN 0 85109 130 X) E. HANDSCOMBE

Price per book for the Technological Series **25s.—£1.25 net** *in U.K. only*